EXTREMES

Humanity is confronted by and attracted to extremes. Extreme events shape our thinking, feeling, and actions; they echo in our politics, media, literature, and science. We often associate extremes with crises, disasters, and risks to be averted, yet extremes also have the potential to lead us towards new horizons.

Featuring essays by leading intellectuals and public figures arising from the 2017 Darwin College Lectures, this volume explores 'extreme' events, from the election of President Trump, the rise of populism, and the Brexit referendum, to the 2008 financial crisis, the Syrian war, and climate change. It also celebrates 'extreme' achievements in the realms of health, exploration, and scientific discovery.

A fascinating, engaging, and timely collection of essays by renowned scholars, journalists, and intellectuals, *Extremes* challenges our understanding of what is normal and what is truly extreme, and sheds light on some of the issues facing humanity in the twenty-first century.

Contributors

David Runciman, Emily Shuckburgh, Nassim Nicholas Taleb, Roz Savage, Lyse Doucet, Matthew Goodwin, Sarah Harper, Andrew C. Fabian

DUNCAN NEEDHAM is Dean and Senior Tutor of Darwin College, Director of the Centre for Financial History, a Senior Researcher at the Centre for Risk Studies, and an Associate Lecturer at the Faculty of History, University of Cambridge.

JULIUS WEITZDÖRFER is Director of Studies in Law and a former Charles and Katharine Darwin Research Fellow at Darwin College, a Research Associate at the Centre for the Study of Existential Risk, and an Affiliated Lecturer at the Faculty of Law, University of Cambridge.

THE DARWIN COLLEGE LECTURES

These essays are developed from the 2017 Darwin College Lecture Series. For more than 30 years now, these popular Cambridge talks have taken a single theme each year. Internationally distinguished scholars, skilled as popularisers, address the theme from the point of view of eight different arts and sciences disciplines.
  Subjects covered in the series include

Extremes

Edited by *Duncan Needham* and *Julius Weitzdörfer*

University of Cambridge

CAMBRIDGE
UNIVERSITY PRESS

CAMBRIDGE
UNIVERSITY PRESS

University Printing House, Cambridge CB2 8BS, United Kingdom

One Liberty Plaza, 20th Floor, New York, NY 10006, USA

477 Williamstown Road, Port Melbourne, VIC 3207, Australia

314–321, 3rd Floor, Plot 3, Splendor Forum, Jasola District Centre, New Delhi – 110025, India

79 Anson Road, #06–04/06, Singapore 079906

Cambridge University Press is part of the University of Cambridge.

It furthers the University's mission by disseminating knowledge in the pursuit of education, learning, and research at the highest international levels of excellence.

www.cambridge.org
Information on this title: www.cambridge.org/9781108457002
DOI: 10.1017/9781108654722

First published 2019

Printed and bound in Great Britain by Clays Ltd, Elcograf S.p.A.

A catalogue record for this publication is available from the British Library.

ISBN 978-1-108-45700-2 Paperback

Contents

Figures

Figures

Tables

Notes on Contributors

Lyse Doucet OBE is the BBC's award-winning Chief International Correspondent, who spends much of her time covering stories in our news headlines, including devastating wars in Syria and Iraq as well as Afghanistan. She often focuses on the human costs of conflict. Her work also involves asking questions of world leaders. Her BBC journalism began with postings in Abidjan, Kabul, Islamabad, Tehran, Amman, and Jerusalem. She was awarded an OBE in the Queen's Honours list in 2014 for her services to broadcasting and the Columbia Journalism Award for lifetime achievement in 2016.

Andrew C. Fabian OBE is the former Director of the Institute of Astronomy at the University of Cambridge. He also leads the X-Ray research group within the Institute. The group's research focuses of active galaxies, clusters of galaxies, elliptical galaxies, galactic black holes, neutron stars, and the X-ray background. He is one of two UK members of the European Space Agency's Athena Science Study Team. Before becoming its Director, Professor Fabian was a Royal Society Research Professor in the Institute of Astronomy. Formerly, he was President of the Royal Astronomical Society and Vice-Master of Darwin College. He was made a Fellow of the Royal Society in 1996, a Foreign Associate of the US National Academy of Sciences in 2016, and he was awarded the American Astronomical Society's (AAS) Bruno Rossi Prize, the Dannie Heineman Prize for Astrophysics of the AAS and APS, and the Catherine Wolfe Bruce Gold Medal of the Astronomical Society of the Pacific. Asteroid 25157 Fabian was named for him in 2016. Inter alia, he is a co-editor of *Frontiers of X-Ray Astronomy* and of the Darwin Lecture Series volumes on *Origins, Evolution, Conflict, Darwin,* and *Life.*

Matthew Goodwin is Professor of Politics and International Relations at Rutherford College, University of Kent, and Senior Visiting Fellow at

Chatham House. In 2014 he received the Richard Rose Prize and the Communicator Prize for his dissemination of social science research to a wider audience. His books include *Revolt on the Right: Explaining Public Support for the Radical Right in Britain*, which was awarded the Paddy Power Political Book of the Year, and *Brexit: Why Britain Voted to Leave the European Union*.

Sarah Harper CBE is Professor of Gerontology at the University of Oxford and the Founding Director of the Oxford Institute of Population Ageing. Sarah served on the Council for Science and Technology, which advises the Prime Minister on the scientific evidence for strategic policies and frameworks. In 2017 she served as Director of the Royal Institution of Great Britain. Sarah is a Director and Trustee of the UK Research Integrity Office and a member of the Board of Health Data Research UK. Her latest book, *How Population Change Will Transform Our World*, is published by Oxford University Press. Sarah is a Fellow of the Royal Anthropology Institute and she was appointed a CBE in 2018 for services to Demography.

David Runciman is Professor of Politics and Head of the Department of Politics and International Studies (POLIS) at the University of Cambridge. His books include *The Confidence Trap: A History of Democracy in Crisis from World War I to the Present* and *How Democracy Ends*, and he writes regularly about politics for the *London Review of Books*. He is one of the directors of a major Leverhulme-funded research project based in Cambridge on Conspiracy and Democracy, which explores the history and impact of conspiracy theories on democratic politics. He is also the host of the popular weekly podcast *Talking Politics*.

Rosalind 'Roz' Savage MBE, an Oxford-educated lawyer, spent the first 11 years of her career working as a management consultant before an environmental epiphany led to her transformation into a world-class adventurer. In the years between 2005 and 2011 Roz rowed, solo, across the Atlantic, Pacific, and Indian Oceans, making her the world's foremost female ocean rower. On the ocean, Roz had to redefine her comfort zone on a daily basis, and reach deep into her inner sources of strength, self-discipline, and commitment to her goal, as she spent up to 5 months alone at sea on a 23-foot rowboat, thousands of miles from land and humanity, at the mercy of winds, waves, and currents. Roz is a Fellow of the Royal Geographical Society and a Senior Fellow at Yale's Jackson Institute, where she has taught a course on Courage. She has written two books about her ocean adventures, *Rowing the*

Atlantic and *Stop Drifting Start Rowing*, and she continues to be an advocate for sustainability.

Emily Shuckburgh OBE is a climate scientist and Deputy Head of the British Antarctic Survey's Polar Oceans Team, which focuses on understanding the role of the polar oceans in the global climate system. She holds a doctorate from and a number of positions at the University of Cambridge and is a Fellow of Darwin College. She has worked at Ecole Normal Supérieure and at the Massachusetts Institute of Technology. She is a fellow of the Royal Meteorological Society, where she co-chairs the Climate Science Communications Group, a trustee of the Campaign for Science and Engineering, and a member of the Scientific Steering Committee of the Isaac Newton Institute for Mathematical Sciences. Emily has also acted as an advisor to the UK government on behalf of the Natural Environment Research Council. She co-authored the book *Climate Change* with the Prince of Wales and Tony Juniper, and the Darwin Lecture Series volume on *Survival.*

Nassim Nicholas Taleb spent 21 years as a risk taker before becoming a researcher in philosophical, mathematical, and (mostly) practical problems with probability. He is the author of a multivolume essay, the *Incerto* (*The Black Swan, Fooled by Randomness, Skin in the Game, The Bed of Procrustes: Philosophical and Practical Aphorisms*, and *Antifragile*), exploring the broad facets of uncertainty. His books have been translated into 36 languages. In addition to his life as a trader, Taleb has published, as a backup to the *Incerto*, more than 45 scholarly papers in statistical physics, statistics, philosophy, ethics, economics, international affairs, and quantitative finance, all on the notion of risk and probability. He has spent time as a professional researcher (Distinguished Professor of Risk Engineering at New York University's School of Engineering and Dean's Professor at the University of Massachusetts, Amherst).

Acknowledgements

This book would not have come together without the Darwin College Lectures on 'Extremes', on which the following essays are based, and without the many members of the College who facilitated the eight-week event in 2017. The annual lecture series was inspired by Andy Fabian more than thirty years ago, and under his stewardship this and all preceding series have come together successfully not only as the annual flagship event of Darwin College, but also as some of the best-attended events in the Cambridge calendar.

We are *extremely* fortunate, therefore, in the number of College staff and fellows who have been helping with the lectures. The Education and Research Committee, chaired by Andy Fabian, offered advice when we curated the 'Extremes' series. Janet Gibson made the necessary arrangements for speakers and guests and successfully publicised the lectures, as every year. Espen Koht, Jamie Pilmer, Markus Kalberer, Tony Cox, and the student camera crew of Meredith Hadfield and Michael O'Neill ensured that light and sound in lecture theatres and the video recordings went smoothly, while Roger Whitehead and his team ushered our up to 700 attendees per evening, of whom as many as possible were then entertained thanks to our talented catering team. Finally, there were guests to be introduced and thanked by our Master, Mary Fowler.

First and foremost, Janet Gibson's patience, dedication, and friendly and meticulous organisation have been central to this book from beginning to end. Sarah King provided draft transcripts of some of the lectures. The cover photograph, depicting three of the twenty-three wards of the megacity of Tokyo (Taitô-ku, Sumida-ku, and Chiyoda-ku), is used by courtesy of Darwin alumnus Timur Alexandrov. We are grateful to Libby Haynes at Cambridge University Press for her help in editing this collection, and especially grateful to David Runciman for stepping in at short notice when Prime Minister Theresa May dropped out.

Julius Weitzdörfer expresses his heartfelt gratitude to Martin Rees, Huw Price, Seán Ó hÉigeartaigh, Catherine Rhodes, and to the Templeton World Charity Foundation for support during his work on this book, in the framework of the project 'Managing Extreme Technological Risks' at the Centre for the Study of Existential Risk.

We dedicate this book to Professor Sir David J.C. MacKay FRS FInstP FICE (1967–2016), polymath, physicist, mathematician, energy and sustainability scientist, Regius Professor of Engineering, Fellow of Darwin College, and Chief Scientific Adviser to the UK Department of Energy and Climate Change.

Duncan Needham & Julius Weitzdörfer

On the Notion of 'Extremes'

JULIUS WEITZDÖRFER

Humanity is confronted by and attracted to extremes. Extreme events shape our thinking, feeling, and actions, and echo in our politics, media, literature, and science. We often associate extremes with crises, disasters, and risks that are to be averted. Yet extremes also have the potential to lead us towards new horizons.

Featuring essays arising from the 2017 Darwin College Lectures, this volume explores a spectrum of 'extreme' events, from the 2008 financial crisis, the election of President Trump, the rise of populism, and the Brexit referendum, to the Syrian war and climate change. At the same time, the essays celebrate 'extreme' achievements in the realms of human health, ocean exploration, and cosmological discoveries, shedding light on extremes in the past, the present, and the future.

In this preface, we explore the notion of extremes by reflecting on the individual contributions to the volume. These essays challenge our understanding of what is 'normal', what is exceptional, and what 'extreme' really is by illustrating how extremes are the manifestation of what we do and what we perceive, and are at the heart of many issues we cope with. Pulling together the core topics of the book, we show how the notion of extremes radiates into different subject areas, and the way in which it can – and cannot – help us make sense of the world.

Three fundamental questions link the essays.

The first is why there is such a human fixation on (if not an obsession with) and such widespread attention to extremes: from setting records to attributing superlatives, from climbing the highest mountain to exploring the poles, from driving the fastest car to baking the best cake, from collecting the rarest stamp to catching the biggest fish, from finding the oldest fossil to detecting outliers within the realm of machine learning. We, the editors, with our respective interests in economic and environmental crises, financial and nuclear meltdowns, historical and future dimensions of risks, admittedly share some of the (academic) excitement around extremes.

This fascination permeates this volume. For instance, driven by excitement and determination, Roz Savage spent up to five months alone at sea, rowing for twelve hours a day. As the first (and so far, only) woman ever to do so, she discusses the challenges of rowing solo around the globe – and the motivational force of extremes – in her essay 'Extreme Rowing'. Our fixation with extremes also makes them a magnet for media attention. In her chapter, Lyse Doucet shares her experiences of reporting on 'Extremes of War' as the BBC's Chief International Correspondent – most recently from Syria. The sometimes fascinating, sometimes frightening 'appeal of extremes' as a challenge, a source of learning, and sometimes as an object of voyeurism can possibly be attributed to human psychology and evolution. So, be it by instinct or by reason, extremes attract our attention – but what is the merit?

The second, perhaps more important question, therefore, revolves around the value of extremes for human knowledge. What is the epistemic value, the intellectual advantage of experiencing and understanding the 'extreme', the anomalous, over the 'normal' and the mundane? Can a dichotomy of 'extreme' versus 'mean' be a useful way of looking at the world? A basic hypothesis about our attraction to extremes could be that we think about them assuming a kind of polarity, attempting to use them as starting points for locating the 'normal' in the middle. However, a focus on fringe or marginal phenomena can also distract from and impair the view of the ordinary, in other words, what seems to matter most, most of the time. Astronomer Royal Martin Rees, who gave the first-ever Darwin Lecture in 1986, suggests that advances in astronomy – the oldest numerical science – offer a clue. Summarising centuries of discovery, and looking into the future, Rees posits,

> Astronomers are always specially interested in the most 'extreme' phenomena in the cosmos, because it is through studying these that we are most likely to learn something fundamentally new... There are three great frontiers in science: the very big, the very small, and the very complex.[1]

In other words, by looking to the edges of what we know, what we can observe, and what we can imagine, we are more likely to discover the fundamental, the new, and the surprises. Often it is in the minimal and maximal limits that our best ideas are tested – and found. As extremes test our understanding, they can cause a crisis of an academic discipline, expose flaws in a theory, and eventually trigger the paradigm shifts that our knowledge systems irregularly undergo. In this way, extremes can constitute 'anomalies', as referred to in the context of scientific revolutions by philosopher of science Thomas Kuhn.[2] Their observation as new phenomena, or construction as novel ideas, has the potential to challenge existing worldviews. Similarly, as Joseph Overton has argued, positions previously conceived as 'extreme' can be, gradually or suddenly, accepted and embraced as 'normal' within public discourse, or in the process and wake of social, economic, and political changes.

Intellectual history, the history of ideas, and the history of science offer abundant examples of how extremes test our deepest-held knowledge. In some cases, resistance to extremes can prevail, e.g. in the case of religious extremism, and, to the contrary, cause pushback and backlash, often towards the opposite 'extreme'. Against this backdrop, Matthew Goodwin's essay 'Extreme Politics' explores the recent resurgence of the far right in Europe. In other cases, formerly 'extreme' positions can become a 'new normal', such as the Lutheran reformation, Copernican heliocentrism, or Wegener's theory of plate tectonics. As another example, in her essay 'Extreme Longevity', gerontologist Sarah Harper explores the profound economic and social consequences that follow from the fact that half of Europeans born today will live to be a hundred years old, and whether there is a maximum life span for any human being. In situations such as this, it is impossible to uphold the status quo ante or restore the status quo post in a way identical to the previous state of affairs. That is, as extreme events generate 'new normals', old patterns are disrupted, and new patterns emerge. Thus, in the process of anomaly-driven shifts, insights, and theory redesigns, extremes have the potential to open the doors to understanding and accepting what we would have never considered valid before.

Even where experiencing, achieving, or studying extremes does not lead to 'fundamentally new' insights, it can provide impetus to incremental adjustments of our existing models of the world. This is especially the case where we understand extremes as lying quantitatively (among a distribution of values) or qualitatively (on a gradual spectrum), instead of classifying them as categorical extremes entirely beyond previous knowledge systems. Wherever the 'merely gradual' or 'merely quantitative' extremes can be accommodated, there is a tendency for them to be subsumed by existing frameworks, simply by moving the metaphorical goalposts between the possible and the impossible. However, insofar as extremes constitute categorical 'outliers', i.e. rare objects causing analytical problems, they are opportunities not only to shift the known limits of sample maxima and sample minima, but also to significantly redefine what constitutes the average and what we henceforth consider as 'normal'.

While in statistical distributions an extreme value can be referred to as a maximum or minimum upwards of four sigmas (σ), Nassim Nicholas Taleb's essay 'Probability, Risk, and Extremes' demonstrates the shortcomings of conventional statistical tools. Taleb demonstrates how supposedly robust statistics are not robust at all, how frequency-based forecasting fails, and how past averages misrepresent future ones. So, regardless of how we conceptually understand deviations, outliers, exceptions, novelties, and other 'abnormal' phenomena as extremes, we can acknowledge their potential and their significance in challenging us, generating new ideas, and stress-testing existing ones. Thus, there is significant scholarly merit in treating them as objects of study, as keys to unlock the treasure chests of knowledge.

Finally, the essays address, explicitly or implicitly, and each in its own way, whether we are living in an age of extremes. This is a compelling proposition when thinking about the unprecedented number of humans on planet Earth and their irreversible footprint on the biosphere. Albeit, words of caution might be warranted. What goes without saying is that ruling out observational and interpretational errors, and awareness of anthropic bias, is essential before jumping to the conclusion that we are observing an 'extreme' phenomenon. To this end, David Runciman's essay 'Dealing with Extremism' teases out the important differences between political extremists and conspiracy theorists: while the experience of governing tends to soften the edges of extremists, conspiracy theorists become hardened in their views, turning conspiracy theories into governing philosophies. Not only in the political sphere, we are also well advised to remain conscious of who gets to define ideas as 'extreme' and to whose benefit this is. Normatively, we do not necessarily know an 'extreme' when we see one as power relationships tend to influence anomaly construction.

Most importantly, therefore, the essays in this book should remind us that the notion of the 'extreme' is fundamentally relative, for the classification or declaration of anything as 'extreme' is context-sensitive and ephemeral: tied to a particular time, place, and perspective. Because yesterday's extremes all too often become today's 'normal', they are contingent on their (explicit and implicit) frames of reference, scale, scope, and limitations of observation. For example, in Emily Shuckburgh's essay 'Extreme Weather' on the science underlying the causes and implications of climate change, we learn that 2016 was the warmest year on Earth since records began, the second warmest being 2015 and the third warmest 2014. Such comparisons are inherently contingent on the timeframe of our records, in this case excluding non-anthropogenic climatic extremes (around 55.5 million years ago temperatures were possibly 8°C warmer globally than today) and our, in all likelihood, hotter future.

The depth of the past, the vastness of the cosmos, the uncertainties of the present, and the unknowns of the future are humbling, and, in conclusion, must caution us to deem anything 'extreme'. Such attributions should therefore only be made, and read, in quotation marks. Andrew Fabian's discussion of the life and death of stars in 'Extremes of Power in the Universe' takes us to the limits of human imagination and thereby perfectly illustrates this point. Across historical, archaeological, palaeontological, geological, or astronomical scales of time, and across space (which, according to multiverse theory, could span across an infinite number of universes), indeed, it is impossible to refer to anything as 'extreme' with certainty.

In summary, the notion of 'extremes' appears to be of value in three ways: when it drives our curiosity and aspiration, when it helps us understand the world, and when it makes us realise our own humility.

'Critical rationalism' and the concept of empirical falsification, famously advanced by Karl Popper, the philosopher of science often compared against Kuhn (and formerly an Honorary Fellow of Darwin College, as is Rees today), suggest that, in the empirical sciences, extremes can never be proved.[3] They can only be falsified by even 'extremer' extremes. English grammar, thus, rather logically, allows higher degrees of the adjective 'extreme'. So as much as extremes remain a mirror of the limitations of their context and time, and of the humility of human knowledge, they should be referred to with caution.

References

1 M. Rees, *Just Six Numbers* (Basic Books, 2000), pp. 35, 159.
2 T. Kuhn, *The Structure of Scientific Revolutions* (University of Chicago Press, 1962).
3 K. Popper, *Logik der Forschung: Zur Erkenntnistheorie der modernen Naturwissenschaft* (Mohr Siebeck, 1934); rewritten as *The Logic of Scientific Discovery* (Hutchinson Education, 1959).

1 Dealing with Extremism

DAVID RUNCIMAN

'Dealing with extremism' is a resonant phrase but one that can mean very different things in different contexts. It is something that is most often said by politicians in the context of national security. A government perspective on dealing with extremism would mean knowing how to isolate it, control it, minimise it, suppress it, where necessary criminalise it, and, if possible, eliminate it. That is one sense of how you 'deal with' something – it is how you deal with your enemies. These are real and serious political challenges and I am not seeking to minimise their importance when I say my perspective is not that one.

How we deal with extremism can raise a very different set of questions about whether and where it is possible to draw the line between extremism and other kinds of dissenting politics and belief. That is what I want to discuss here: the difficulty of knowing what counts as extremism in an age of a widespread loss of faith in established political institutions. Various forms of democratic political dissent share characteristics we might also associate with extremism: a refusal to compromise – indeed, an explicit repudiation of compromise; a suspicion of democratic values; a deep mistrust of conventional sources of power and authority. Democratic governments tend to want to draw as sharp a line as possible between extremism and the rest of political life. But what if the line is blurred and becoming more blurred all the time? How do we deal with that?

These questions have become far more acute since January 2017, with the advent of the presidency of Donald Trump. What makes it increasingly hard to draw the line between extremism and other forms of political life is the fact that some seemingly extreme views have come to occupy a place at the heart of American politics. Donald Trump would have us believe that far from being an extremist, his primary motivation has been to deal with the threat of violent Islamist extremism. But does this distinction hold up? It is partly a question of definitions, and who is doing the defining, and that is something I shall discuss later. But sometimes we do not really need to quibble about definitions. Some of the things that Trump has said – about Islam, about immigrants, about his opponents – are characteristic of a political rhetoric that is normally associated with extremist ideology, including but not limited to forms of white supremacism. Trump as president has retweeted messages from political groups – such as Britain First in the UK – that are routinely described as extremist. If it looks like a duck and sounds like a duck, but somehow manages to waddle its way into the White House, is it no longer a duck? On the other hand, if the person who brings this baggage into the White House has done so by legitimate means, having won an election under the rules of the game (weird though those rules might appear, given that the winner of the popular vote lost the election), how do we deal with that?

Let me use a fictional Mafia example to illustrate the distinction that underpins how I will be interpreting the title of this chapter. If one of Tony Soprano's henchmen were to come to him and whisper in his ear about a problem – a betrayal, a deal gone sour, a threat to the business – and Tony were to say, 'Deal with it,' we would know what that meant. It means make it go away. That is how a Mafia boss deals with his enemies. But dealing with Tony's enemies is only a small part of what *The Sopranos* is about. It is also about how Tony deals with his demons. We cannot draw the line so easily when it comes to his demons, because they are inside him and inside his family. Dealing with his demons is what sends Tony Soprano to a psychiatrist.

One way to explore how we might deal with extremism in this second sense is to look at the relationship between extremism and conspiracy theories, which are among the demons that increasingly haunt contemporary political life. Conspiracy theories are both a symptom and a cause of what Richard Hofstadter memorably labelled the 'paranoid style' of politics and therefore perhaps one of the things that makes our political culture fit for the psychiatrist's couch. Wild conspiracy theories about secret plans and plots inside government, about foreign infiltration and hidden agendas, are a sign of deep distrust of government – and often a manifestation of our underlying fears. But we are also afraid *of* conspiracy theories, perhaps excessively so, seeing them as a sign of 'post-truth' values and the collapse of reason and responsibility.

I will discuss the relationship between extremism and conspiracy theory for two reasons. First, I have been involved in a long-term project that has studied the connection between conspiracy theories and democracy.[1] This has involved trying to address the vexed question of where and how to draw the line between dangerous levels of mistrust on the one hand and unavoidable or even necessary levels of suspicion on the other. Democracy needs routine suspicion – one of the things that distinguishes it as a form of politics is that it does not require the general public to believe what they are told. In a democracy we are all allowed to make up our own minds. Furthermore, we all know that there are some real conspiracies out there – occasionally the wild suspicions that some people have about the infiltration or corruption of democratic life turn out to be true. The idea that Donald Trump is a stooge of the Russian government, which is said to possess secret film of him in compromising situations, is a conspiracy theory. But the idea that the agents of the Russian secret state do attempt to gather and use such material is far from outside the bounds of reasonable suspicion – after all, they have history in this area. Not all conspiracy theories are false. And not all political extremism falls outside of the domain of government action.

That is the other reason for discussing conspiracy theories – they have moved closer to the heart of democratic politics, as shown by the dramatic recent return of the question of whether the Russians are subverting Western democracy. When we embarked on our investigation of conspiracy theories in 2012, at the outset of the Conspiracy and Democracy project, we were studying what was largely taken to be a fringe activity. The questions we were most often asked were, why does it matter if some people believe the craziest things: what harm does it do? But over the past few years the questions have become much more urgent. People now want to know why conspiracy theories are so prevalent and why they are so hard to counter. Moreover, this is not just an American phenomenon, let alone something that is limited to Trump and the Russians. The Scottish independence referendum in 2014 was bedevilled by conspiracy theories on both sides of the argument, as was the Brexit referendum in 2016. The election of Jeremy Corbyn as leader of the British Labour Party has encouraged conspiracy theories from his supporters – many of whom believe that the bad press he gets is the result of MI5 infiltration of the press – as well as his opponents – who see Momentum as a Trojan horse to infiltrate hard-left ideas into mainstream British politics. Conspiracy theories can appear to be everywhere at present.

I want to offer a basic proposition to help see if it is possible to draw the line between conspiracy theory and extremism. My claim is this: *Not all conspiracy theorists are extremists, but almost all extremists are conspiracy theorists.* That is, people who get classified as extremists by those at the centre – and to be an extremist means being pushed to the extremes or margins of mainstream debate, out beyond the edges of what is acceptable – are liable to be conspiracy theorists; but people who are classified as conspiracy theorists are not so sure to be extremists. Let me take each part of that claim in turn before I try to explain the relationship between them.

Why do I say not all conspiracy theorists are extremists? I have two connected reasons. First, there is evidence that believing in conspiracy theories is actually quite widespread. If believing in conspiracy theories makes you an extremist, then we would have to face the fact that there are an awful lot of extremists out there. Take the Corbyn example just mentioned – in a 2016 poll, half of Corbyn's supporters in the Labour Party claimed to believe that MI5 is working to destabilise him and that it is the activities of the secret services that explain the hostile press he receives. On the one hand, that suggests extreme mistrust of the British state and in the basic functioning of democracy. On the other hand, I would not want to call the majority of Corbyn's supporters extremists – I know these people, and from my experience holding this belief actually goes along in many cases with gentle, conciliatory, and rather tolerant political personalities.

We conducted our own surveys for the Conspiracy and Democracy project.[1] Belief in hardcore conspiracy theories is quite low – only a relatively small proportion of the population subscribes to the idea that the moon landings were fake or that Diana was murdered by MI6. But propositions such as 'the EU is planning to take over the laws of this country' or 'the government is deliberately hiding the true immigration figures' are quite high – around half of all voters believed those statements in 2015 (the first question is now a little out of date; the second one is not). Two-thirds believe that whatever happens in elections, 'the same small group of people run things'. If we have to conclude that more than half of all voters are therefore extremists – fundamentally opposed to democratic principles, because they think democracy is a front for the real power relations that control our society – then we have a serious problem. Indeed, we do not have a democracy. So I am disinclined to believe that.

It is also unclear that people neatly divide between those who are conspiracy theorists and those who are not. There is substantial evidence that belief in conspiracy theories is quite fluid and tends to track democratic outcomes rather than standing in opposition to them. In the United States, conspiracy theories among Democrats rise when there is a Republican in the White House, and vice versa. For obvious reasons, Democrats were far more likely than Republicans to believe that George W. Bush was a stooge of the oil industry who put him up to invade Iraq. Likewise, Republicans (including, notoriously, Trump himself) were far more likely than Democrats to believe that Obama was not born in the United States and is in fact a secret Muslim, for equally obvious reasons. When the other side gets in, people on your side are more likely to believe that things are not what they seem; when your side gets in, suddenly you are more reconciled to what is going on. So this could be part of the ebb and flow of democracy rather than a threat to it: a feature of the system, not a bug.

That said, there are some people who believe in conspiratorial explanations regardless of who wins elections. But our research found that these people are often the ones who feel excluded altogether from democratic politics – they do not have a side in the mainstream fight because they feel all possible winners are equally bad. In the UK, for instance, that includes a significant number of hardcore UKIP supporters prior to the Brexit referendum who were sometimes denounced as extremists.[2] But that is not how it looks from this perspective: to believe in EU plots to subvert British democracy may simply be an expression of disgust with a system that seems to hoard power in one place and to be resistant to change.

Belief in conspiracy theories is an outlet for frustration with the system – and indeed from the point of view of the excluded an expression of basic support for democratic values, taken to have been corrupted – rather than a repudiation of democratic values.

This links to the second reason why it is a mistake to think of conspiracy theorists as extremists – many people hold beliefs that posit secret plots and conspiracies that make a mockery of democracy, but they have no intention of acting on these beliefs. In that sense, the beliefs are extreme, but they are not extremist beliefs because they are in no sense a call to arms. They are the opposite – an expression of resignation, of loss of faith in the capacity of political action to change things. There is such a thing as quiescent conspiracy theory. And it could be called uncompromising, in that it is a refusal to engage with the world on its terms, which demand compromise and a willingness to listen to the other side. One way to refuse to engage with the world is to shut the world out, which is what many conspiracy theorists do; but it is not what extremists do. The clichéd image of the conspiracy theorist fits with this view: a man, alone in his basement, obsessively charting his elaborate paranoia on a blinking computer screen. This might come close to the clichéd image of the extremist: another man in a basement, secretly plotting to blow up the world. But they are different, and we should not confuse them.

However, the quiescent kind is not the only sort of conspiracy theory, nor is being resigned to the duplicity of the world the only way to be a conspiracy theorist. There are active varieties as well. So on to the other half of my proposition: almost all extremists are conspiracy theorists. I am taking extremism in this sense to mean the active repudiation of the established order as a system of oppression. When the established order is democratic, that means repudiating democracy. This fits with the British government's official definition of extremism as 'vocal or active opposition to fundamental British values, including democracy and the rule of law . . .' Extremism in this sense invariably involves the demonisation of certain groups as responsible for the oppression and the advocacy of violent or at least extra-constitutional means to resist it. It is what connects Islamist extremism with far-right racist groups. But something else that connects them is that extremists of both these stripes are invariably conspiracy theorists. They point to the hidden forces at work pulling the strings behind the scenes, whether it is the CIA, the federal government, or more broadly, and more depressingly, the Jews.

Of course, definitions of extremism very much vary depending on what is deemed to be a threat to democratic values. When I was a student in the late 1980s and early 1990s, the British government banned Sinn Fein in order to deny what was deemed an extremist organisation 'the oxygen of publicity'. Sinn Fein had relatively little in common with some current groups that fall under the extremist label. Many of the party's political views were much more conventionally mainstream. But one thing they did share was a propensity towards conspiracy theories: that is, to seeing the real story of democracy (in this case, British democracy) as being not what takes place on the surface, but as being determined by the secret actions of the secret state. And that is partly because the secret actions of the secret state were what determined Sinn Fein's experience of British democracy.

In other words, extremists may have good reason for thinking that democracy is a front for something else, precisely because they have been excluded from it. The talk of openness and tolerance is not for them. That necessarily implies that the real story is known only to those behind the scenes, with access to the secret world of power and connections. So the official version is to be treated as nothing but a cover-up. Let me be clear about what I am not saying. I am not saying that extremists are paranoid simply because the state is out to get them. Some extremists – including some current Islamist or far-right extremists – are deeply paranoid as part of their worldview, which is based on long-standing and violent narratives of betrayal and conspiracy. Being labelled an extremist does not make you an anti-Semite. Most of these people are already anti-Semites. And many of those who are the nastiest kind of anti-Semite are the ones who get labelled extremists. But even extremists who are not anti-Semites can still wind up with a very conspiratorial view of politics, because of how they experience it.

There is another way to put this. One reason extremists are conspiracy theorists is that the active rejection of democratic values requires secret organisations and conspiratorial behaviour: to be an extremist you have to be a conspiracist, in the sense that you have to cover up your activities, to keep them away from the prying eyes of your enemies, above all the state. One reason that extremists think the world is out to get them is that the world really is out to get them. Sinn Fein suspected the British state of conspiring against them and they were right. There is considerable evidence that people who engage in conspiracies also tend to see the world in conspiratorial terms: that is, they imagine that their enemies behave like them. The paranoia of Stalinism derived in part from Stalin's early experiences of persecution and the need for secrecy: leading a secret life will attune almost anyone to seeing the world as a series of secret undertakings. To plot violence invariably leads you to believe that others are plotting violence against you. And often they are (as they certainly were against Stalin, only on nothing like the scale he eventually imagined). So these states of mind become mutually reinforcing: extremists have their worldview confirmed by the fact that the world treats them as dangerous extremists. They become conspiracy theorists because they know that conspiracies are real.

My argument therefore is that believing in conspiracy theories and being an extremist are not the same thing. But the two do overlap. Almost all extremists happen to be conspiracy theorists, sometimes by conviction and sometimes by circumstance. And some conspiracy theorists happen to be extremists, but by no means all. What follows from this? Let me suggest a number of possible implications, using some examples from contemporary politics. First, we must be careful not to read extremism off conspiracy theories, rather than conspiracy theories off extremism. In other words, we must be careful not to brand all radical rejections of mainstream politics as extremist. Nick Clegg, when Deputy Prime Minister, in a speech in 2011 setting out his view of how to deal with extremism, characterised its primary failing as being a product of 'closed minds'. But most of us have closed minds on some issues – and certainly conspiracy theorists do. It does not make us all extremists. We also must remember that for those who feel excluded from democracy because the system never seems to take their views seriously, the closed-mindedness is all on the other side. From this perspective, the political establishment is where the extremism lies.

Dominic Cummings, in a blog post from early 2017 that recounts his understanding of the causes of the Brexit vote that he helped to engineer, makes this point very clearly:

> *It doesn't occur to SW1 and the media that outside London their general
> outlook is seen as extreme. Have an immigration policy that guarantees free
> movement rights even for murderers, so we cannot deport them or keep them
> locked up after they are released? Extreme. Have open doors to the EU and don't
> build the infrastructure needed? Extreme . . . Ignore warnings about the dangers
> of financial derivatives, including from the most successful investor in the history
> of the world, and just keep pocketing the taxes from the banks and spending your
> time on trivia rather than possible disasters? Extreme. Make us – living on
> average wages without all your lucky advantages – pay for your bailouts while
> you keep getting raises and bonuses? Extreme and stupid – and contemptible.
> These views are held across educational lines, across party lines, and across class
> lines. Cameron, Blair, and Evan Davis agree about lots of these things and tell
> people constantly why they are wrong to think differently but to millions they are
> the extremists.*[3]

We should therefore be very careful who we call an extremist. People
who reject mainstream democratic politics are often doing so because
they think mainstream democratic politics has rejected them. It does not
follow that they do not believe in democracy. In fact, they may feel they
are the only ones who do still believe in it.

Certain forms of political expression and belief – the search for hidden
explanations behind the official version of events; the tendency to scape-
goat particular groups or institutions; the feeling that democracy is a fix;
a deep reluctance to accept what is being presented as 'the truth' – are
shared by both extremists and conspiracy theorists. It is wrong to
conflate them: it is possible to hold these sorts of views in very different
ways. Yet there is a real danger that we have increasingly begun to
conflate them. That is because, as I suggested earlier, the image we have
of the conspiracy theorist is of the lone wolf, somewhere on the edges of
society, very much in the minority (sometimes in a minority of one): the
renegade, capable of blowing a fuse. To a lot of people, that sounds like
the definition of an extremist. But many conspiracy theorists are too busy
pursuing their research down the rabbit holes of the Internet to pose
much of a threat to anyone. Most genuine lone wolves are not extremists
because they lack the support network to turn their beliefs into actions.

Part of the confusion is that 'conspiracy theorist', like 'extremist', has
become a term of abuse. In that sense, what counts as being a conspiracy
theorist is very much in the eye of the beholder. No one likes to be called

one, and politicians increasingly use the term to stigmatise their most persistent critics. For example, when under pressure, Tony Blair had a tendency to call the people who went furthest in their criticism of the Iraq war 'the usual conspiracy theorists'. By this he meant the people who said it was all a secret plot between him and Bush to steal the oil. By calling them conspiracy theorists he was implying (i) there was no point arguing with them, so he wouldn't bother, and (ii) they were confusing mistakes with malign intent, or, to put it in more familiar terms, they had mistaken 'cock-ups' for 'conspiracies'. I happen to think Blair was right about this and that incompetence rather than conspiracy should carry most of the burden of blame for the failure of the Iraq war (which does not mean I think there was no case to answer about why Blair did not know what he was doing).

However, we should not draw a clean line between conspiracies and cock-ups. These are not mutually exclusive explanatory frameworks. A lot of conspiracies turn into cock-ups: that is the plot of many Hollywood bank heist movies, when a plot goes horribly wrong. More important, a lot of conspiracies are the result of cock-ups: they are attempts to cover up after the event evidence not of nefariousness but of incompetence. Much more likely than that the British state knew in advance that Iraq did not possess weapons of mass destruction is that when the British state discovered that to be the case, against its own expectations, it did a certain amount to cover its tracks. On this account the conspiracy was not the planning for the Iraq war; it was the attempt to conceal the lack of planning. Ironically, the classic example of this sort of confusion may well be the Kennedy assassination, which is normally seen as the ground zero of conspiracy theories. The FBI and the CIA knew all about Lee Harvey Oswald – because Oswald had been a one-time defector to the Soviet Union, they had at various times shown a keen interest in his activities – and after he killed the president both organisations tried to suppress the evidence they possessed of this prior knowledge. Did they do this because they had known he was going to kill the president, and had either turned a blind eye or perhaps even encouraged it? Or was it because they hadn't known, despite having him under surveillance, and so did what they could to cover up their own failures of foreknowledge? To my mind, the second is far more likely than the first.

That is a long way of saying we often stigmatise conspiracy theorists as crazy, and sometimes they are crazy, but more often they are simply vastly over-egging their suspicions, which might otherwise be persuasive. There is no clear dividing line between conspiracy theory and reasonable scepticism, even though it is often possible to tell the two apart. So, in fact, I am using conspiracy theory in a much more expansive sense, to cover a spectrum that runs from the widespread belief that democracy is in the grip of small groups of unaccountable individuals at one end, through to the extremist rejection of basic democratic values at the other. Along this spectrum it is possible to distinguish three broadly distinct manifestations of the different kinds of suspicions I have discussed:

1. There is the suspicion that is part of the ebb and flow of democratic life and which comes and goes as the people you believe in or feel attached to come and go from power.
2. There is the suspicion that becomes entrenched and manifests as a sterile and obsessively mistrustful mindset, which sees hidden plots everywhere and can offer an alternative explanation for every official explanation.
3. There is the suspicion that goes along with an active rejection of democratic norms and manifests as the intention to overturn them.

The name for the first is being a citizen; the name for the second is being a conspiracy theorist; the name for the third is being an extremist. They are different, but they have features in common and they can sometimes look the same.

For example, a belief that Obama was not born in the United States might be evidence that someone is channelling their frustration with the policies of a Democratic administration; alternatively, it might be evidence that such a person is unwilling to accept any official documents (such as Obama's birth certificate) as evidence of anything except the ability of powerful people to cover up the truth; finally, it might be evidence of a propensity to resist the legitimacy of a non-white president of the United States, with violence if necessary. Angry citizen – obsessive conspiracy theorist – dangerous extremist. There are many more of the first than the second (in July 2016, 72 per cent of Republican voters were reported to have continuing doubts about Obama's citizenship); and many more of the second than the third. Nonetheless, they all believe the same thing: that Obama is not really an American.

One frequently asked question about this sort of radical suspicion is, Do they *really* believe it? That is, are these deeply held beliefs that result in particular political commitments, or are they pre-existing political commitments that result in conveniently held beliefs? In many cases, the second is more prevalent than the first – the beliefs follow from the politics rather than the other way around. But we should not think this is something peculiar either to conspiracy theorists or to extremists. It is something both have in common with the broader category of citizens. Recent research in political psychology provides clear evidence that individuals will tailor their epistemological commitments – what they hold to be true – according to their prior political preferences. If you present a group containing both committed environmentalists and committed climate sceptics with the same set of 'facts' about global warming, they will end up further apart than when they started (and more likely to believe conspiracy theories about where the facts came from). Each reads the evidence in the light of what they are already committed to believing.

We are, in the words of the psychologist Jonathan Haidt, not rational creatures but rationalising creatures – we retrofit our beliefs to suit our preferences, just as a lawyer builds a case to suit whoever happens to be his or her client. A lawyer will be careful not to make too many epistemological commitments – Tony Soprano's attorney does not want to know whether what he is told by his client is true, because he knows that would fatally hamper his ability to make a convincing case. But though we think like lawyers, we do not behave like lawyers, because we are far more reckless with our epistemological commitments. We end up believing the stories we tell ourselves about what we want to be true. Ultimately, we are our own clients, which leaves us deeply compromised. No one is immune from being a prejudicial witness when it comes to their own defence, not even, or perhaps especially not, Mafia bosses. That is why as well as needing a good – that is, an unscrupulous – lawyer, Tony Soprano also needs a good – that is, a scrupulous – psychiatrist.

Which brings me back to the president of the United States. Trump's election campaign saw the widespread dissemination of claims about American democracy that certainly fall under the heading of conspiracy theory. Two-thirds of Trump's supporters in the Republican primaries were still ready to accept that Obama is secretly a Muslim (which goes well beyond a belief that he might have been born outside of the United States, since it would require a much more extensive and deliberate campaign of deception). That is a conspiracy theory. But what kind? I do not think it makes sense to call it extremist, in the sense that it does not, except in a few, rare cases, go along with a commitment to taking violent action against democracy. Most of the people who hold this belief are nonetheless law-abiding citizens: what they are doing is not rejecting democracy; they are simply grumbling about it, unpleasantly. Nor do the majority of those who hold this belief fit the model of the conspiracy theorists who reject all forms of received knowledge or who pursue independent research to discover their own truth. Some Trump support-ers will have looked hard into the questions of Obama's origins and found what they wanted to find, but most will simply be repeating something they have been told or have picked up without even being aware of where the story came from. As well as being relatively conveniently held, a lot of these beliefs are, for want of a better word, lazily held. Lazy politics is less of a threat to democracy than active repudiation.

I am reluctant to leave it at that, however. The current prevalence of conspiracy theories cannot simply be put down to the ebb and flow of democratic life, and it would be a mistake to assume that with Trump in the White House and Obama gone from it, the impetus will go out of this kind of conspiratorial suspicion. Something has fundamentally changed with Trump. Even though the claim that Obama is not an American is not extremism per se, and it is not the worst kind of conspiracy theory, it is not something we should simply accept with a resigned shrug and say, 'That's democracy!' It's democracy all right – it is what our democracy has increasingly become – but it still poses a threat.

There are three reasons for this. First, technology has transformed the way in which the dissemination of information ebbs and flows. Once, it might have been the case that a change in the White House changed the balance of news, as the focus of suspicion turned from one side to the other – in the way that in the UK the BBC is seen as anti-Labour when Labour is in power and anti-Tory when the Tories are in power. Now, however, we can increasingly find the news that suits us regardless of who is in power, and social media platforms tailor our news feeds to reinforce our preferences. So, in this country, both Labour supporters and

Conservative supporters currently think the BBC is against them. And in the United States, Trump's elevation to the White House has not simply meant that the scrutiny is now all on him, because anyone uncomfortable with that scrutiny can choose news sources that do not pursue it. Many Trump supporters get their news from channels that continue to focus on the deceptions of his enemies. This state of affairs sometimes comes under the heading of 'fake news' or a 'post-truth' media environment. Again, I would hesitate before drawing a clear line here. We have not suddenly entered a world that never existed before. There has always been a certain amount of fake news – just read local newspapers from the last great populist explosion of the 1890s in the United States, when the truth seemed as malleable as it is now. The golden age of political truth-telling never existed. But what is different now is the ease with which alternative points of view can be evaded. News was once quite hard to come by, and beggars can't be choosers. We now live in an age of information excess, which leaves us free to luxuriate in our prejudices.

Second, much contemporary suspicion of democracy is not a suspicion of this or that party; it is suspicion of democracy itself and its ability to deliver tangible results for ordinary citizens. That was what Trump was channelling in his campaign, and what he reiterated in his inaugural address – his suspicions were directed against the very institutions through which he now must govern, including his own party in Congress, which he has derided almost as regularly as he has derided the Democrats. Having won the election under the rules, he has continued to cast doubt on the electoral process that delivered him victory. This is, in that sense, ecumenical conspiracy theorising and certainly Trump is an ecumenical figure, neither loyally Republican nor Democrat, because he is equally contemptuous of both, having also been identified as both. He is an echo of the old joke – How can you call me prejudiced? I hate *everybody*. Suspicion on this scale seems to go beyond the rhythms of democratic life. It is not calibrated to anything, certainly not to the electoral fortunes of political parties. It is extreme in that it leaves people nowhere to go to restore their faith or at least to temper their suspicions. The democratic representation of these views requires that contempt for democratic compromise remains front and centre. Extremism is uncompromising, and this is an uncompromising suspicion.

But the big difference is that this uncompromising suspicion is now coming from the seat of power. The idea that conspiracy theories are not dangerous because they are, in the words of Joseph Uscinski and Joseph Parent, two scholars of American conspiracy theory, for 'losers' – i.e. for those who are on the outside of political power – becomes a lot less persuasive when conspiracy theory wins. Trump has turned conspiracy theorising into a governing philosophy, or at least he threatens to do so – using it as a means of avoiding fixed commitments, of sowing dissension and confusion among his opponents, and of harnessing the support of the most disenchanted citizens, out of whom he has managed to assemble a coalition. There are other parts of the world where this has already happened – in Putin's Russia, in Erdoğan's Turkey, in Poland, where the ruling Law and Justice Party uses conspiratorial explanations as its default for explaining anything of which it does not approve. It's not our fault, these politicians say: it's the Russians; it's the Islamists; it's the Jews; it's the foreigners. These regimes are increasingly extreme in the positions they adopt: they are trampling over democratic norms. That is not because they are run by extremists (Erdoğan, for instance, started out as a relatively mainstream Turkish politician). It is because they are run by conspiracy theorists.

So, finally, should we be willing to call this sort of politics, and Trump himself, extremist? Yes and no. No, because it is a governing philosophy, and in that sense has limits at the edges of what makes government possible. It is authoritarian rather than a rejection of claims to legitimate authority. However, we are in uncharted territory. I don't think we know where to map the edges of what makes government possible at present. That is, we don't know how far this way of conducting politics can be pushed within a functioning democracy. And we don't know what will happen in Trump's case when this governing philosophy encounters a crisis or an emergency with which it doesn't know how to deal. Conspiracy theorising is usually sterile because it is the mode of expression of people who feel powerless. When people with real power think in this way, there is a risk that under pressure they will act on their beliefs. To act on such beliefs is to be an extremist.

Here I would distinguish between Trump and certain other kinds of accidental extremist. Trump is what happens when a conspiracy theorist comes to power. By contrast, when people who have been deemed extremists by established power holders enter power themselves, the conspiracy theories often fall away. They were conspiracy theorists because they were extremists – once they no longer count as extremists, there is no need for them to be conspiracy theorists. In government, Sinn Fein have shown themselves capable of recognising the limits of democracy, as have other one-time extremists' movements that have entered power. This contrasts with those who are conspiracy theorists through and through. I do not think Trump is an accidental conspiracy theorist. Trump started out as a conspiracy theorist, not as an extremist. Because that is his philosophy, there is no reason to believe that in government he will not act on it. If there is a terrorist attack on US soil under his presidency, or an economic collapse, or a natural disaster, he may seek to blame the hidden powers that caused the disaster and to pursue them to the limits of the law and beyond. He may not get away with it, because the rule of law still operates in the United States. But we don't know for sure.

Extremism in the sense that the British government uses the term – incitement to terrorism and the violent rejection of democratic values – is real and it needs to be combatted as best we can.[4] I do not deny that for a moment. My argument here, though, is that that does not limit the ways that extremism has come to infiltrate our political consciousness and to colour our political imaginations. That raises other dangers. One is that we are too cavalier with how we use the term and what we label extremism. Because extremist beliefs overlap with more widespread suspicion of democracy, some of which is equally uncompromising but much less dangerous – it is what I have called sterile conspiracy theory – we shouldn't think they are conterminous. Believing Obama is a Muslim does not make you an extremist, any more than believing the EU is a secret plot to take over the laws of this country makes you an extremist. Further, many of these beliefs overlap with more routine democratic disenchantment, which ebbs and flows with the fortunes of life in a free society. Believing that regardless of who wins elections, a small group of people still runs things doesn't make you a conspiracy theorist in the pejorative sense: it may mean that you are a realist. So I would agree with Dominic Cummings that we should be very careful about calling Leave voters extremists, even the ones whose views are shaped by a deep suspicion of certain groups or interests. We should be equally careful about calling Trump supporters extremists, even those who make unpleasant noises about America having fallen into the wrong hands. From their point of view, the people who call them extremists are the extremists. That is wrong too. But two wrongs don't make a right.

However, this argument also runs the other way. Being a conspiracy theorist does not make you an extremist. But it makes extremism possible. Because this is a spectrum of belief, and there are no clear dividing lines, we also need to be wary of assuming that the worst beliefs can be isolated or that we can be inoculated from them. We can't. In some circumstances, suspicion of democracy fuels the rejection of democracy, and rejection of democracy can fuel the erosion of democracy. That is happening in Turkey. It is starting to happen in Poland. It could happen in the United States. Which means we shouldn't assume that it couldn't happen elsewhere too. We should be worried about the spread of conspiracy theories and the readiness with which they can be assimilated into political argument, particularly in the age of Facebook and 'fake news'. We should be especially wary of what happens when conspiracy theories spread to the heart of government, because that gives people in government the power to act on them.

Further Reading

P. Mishra, *Age of Anger: A History of the Present* (Allen Lane, 2017).

J.-W. Müller, *What Is Populism?* (Penguin Books, 2017).

K. Olmsted, *Real Enemies: Conspiracy Theories and American Democracy, World War I to 9/11* (Oxford University Press, 2009).

D. Runciman, *The Confidence Trap: A History of Democracy in Crisis from World War I to the Present* (Princeton University Press, 2013).
How Democracy Ends (Profile Books, 2018).

Notes and References

1 Centre for Research in the Arts, Social Sciences and Humanities, *Conspiracy and Democracy: History, Political Theory, Internet*, www.conspiracyanddemocracy.org (accessed 24 July 2018).

2 On the Brexit referendum, see also Chapter 6, 'Extreme Politics: The Four Waves of National Populism in the West' (the editors).

3 On the EU referendum #21: Branching histories of the 2016 referendum and the 'frogs before the storm', http://dominiccummings.com/on-the-eu-referendum (accessed 24 July 2018).

4 For an essay on these challenges from The Darwin Lectures, see L. Zedner, 'Terrorism and counterterrorism', in L. Skinns, M. Scott, and T. Cox (eds.), *Risk* (Cambridge University Press, 2011), pp. 109–130 (the editors).

2 Extreme Weather

EMILY SHUCKBURGH

The lecture from which this chapter is derived took place on Friday, 20 January 2017, as it had just been announced that through 2016, for the third year in a row, the Earth had experienced record warm temperatures. The lecture also coincided with the inauguration of Donald Trump,[1] someone who had publicly questioned the reality of anthropogenic climate change, as president of the world's second-largest emitter of greenhouse gases.

Extreme weather seems to be characteristic of our time, with meteorological records being broken again and again and repeated stories of the misery brought by storms, flooding, heat waves, droughts, and wildfires across the world. The overarching question I want to address in this chapter is whether we are at risk of more extreme weather over the coming decades as our climate changes and, if so, what we can do about it.

Observed Changes to Our Weather and Climate

That the Earth's climate *is* changing is beyond doubt. Records from thousands of weather stations across the world, and ocean data from ships and buoys, show the temperature measured at the Earth's surface has increased substantially over the past century, and especially over the past 50 years. The global average temperature (Figure 2.1(a)) is now more than 1°C warmer than in the pre-industrial era. Changes in many other features of our climate, such as global sea levels, snow cover, Arctic sea ice, and glaciers around the world, have also been observed.

The oceans have been warming not only at the surface but also at depth. As the water warms, it expands to fill a larger volume. This expansion, combined with input of water from melting ice from glaciers

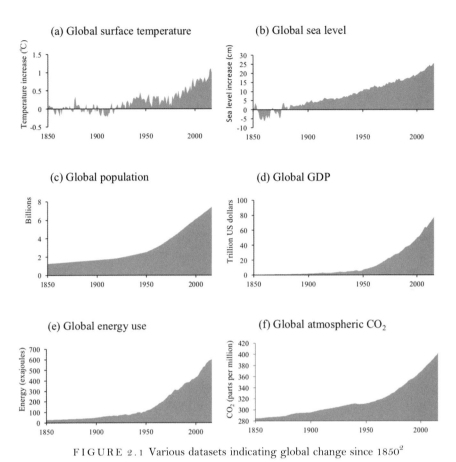

F I G U R E 2.1 Various datasets indicating global change since 1850[2]

and the polar ice sheets, has led to the seas rising by more than 20 cm over the past 150 years (Figure 2.1(b)). About 90 per cent of the additional heat that has accumulated over recent decades has gone into the oceans. It takes a very long time for the oceans to come into thermal equilibrium with the atmosphere, partly because the overturning circulation of the oceans is so slow. The water in some regions of the ocean last saw the surface many hundreds, and in some cases thousands, of years ago. Together with the slow response time of the cryosphere – the frozen parts of the world – this means that we anticipate sea levels will continue to rise for many centuries into the future based on the atmospheric warming we have already experienced.

One of the most striking aspects of climate change has been the rapid decline in the extent of sea ice in the Arctic. For nearly forty years, this has been measured from space by satellite instruments. Consistent with the record surface temperatures, 2016 saw record-low Arctic sea ice extent, with every month of the year ranked among the smallest extents in this satellite record. The change is so substantial that the area covered by sea ice at the end of the summer melt season is now about 2 million square kilometres less than at the end of the twentieth century – a difference equivalent to the combined area of the UK, Ireland, France, Spain, Germany, and Italy. That scale of change has implications for the whole hemisphere, with potential knock-on impacts on weather systems and surface temperatures across Europe, Asia, and North America.

The climatic changes have not been felt uniformly across the world. Several of the feedback processes vary regionally. Some of those – for example, involving sea ice – are unique to the Arctic region and this has meant the Arctic has warmed substantially more than elsewhere, with observed warming about twice that of the global average. Warming-induced reductions in sea ice coverage lead both to less of the Sun's energy being reflected – sea ice being much more reflective than the ocean surface – and to more heat loss from the relatively warm ocean to the atmosphere above it, both of which lead to further Arctic warming.

The personal testimonies of people living in the Arctic bring the data to life and make it very clear that the climatic changes are affecting real people's lives and will continue to affect real people's lives into the future. In 2012, I visited Iqaluit in the Canadian Arctic. There I met a wonderful woman, Mary Ellen Thomas, from the Nunavut Research Institute who described her own interactions with the changing Arctic environment. 'Change is all around us; we are living it,' she said. She described feeling disconcerted by her perception of changing local wildlife, in particular noting that for the first time she had heard a robin singing from the roof of her home. Rather poignantly, she summed up her experience: 'It is as if a friend that we could trust is suddenly acting strangely.'

Around the world, many terrestrial, freshwater, and marine species have shifted their seasonal activities, geographic ranges, migration patterns, abundances, and species interactions in response to ongoing climate change. There is a very fine synchronicity of many species linked to the timing of seasons, and this can be entirely displaced because of climatic changes. Globally, there has been a general pattern of species moving towards the poles or up mountains to remain in favourable climatic conditions. Under natural conditions, many plant and animal species would be able to cope with gradual environmental changes and adapt accordingly. However, rapid climate changes along with the loss of natural habitats, mainly to agriculture and urbanisation, create unprecedented challenges. For species that are already rare, declining, or very specialised, further climatic changes could lead to their extinction. Many of our medicines and all the plants and animals we rely upon for food are derived from wild species. As the rate of extinction continues to gather pace, future generations are likely to inherit a world that lacks the incredible wildlife diversity that we enjoy today and be the poorer for it.

The wildlife in the polar regions is particularly vulnerable to climate change, not least because the options for moving are limited for species that are already residing in the coldest parts of the world. In addition, many of the polar species are incredibly well adapted to the very cold temperatures, but that makes them very sensitive to warming conditions. Cold waters have higher oxygen levels than warmer waters. This allows the survival of Antarctic ice fish (*Channichthyidae*), which have no red blood cells. It also assists the many marine species, such as sea spiders (*Pycnognoida*), that take gigantic forms[3] and hence have a relatively small surface-area-to-volume ratio that is less efficient at supplying tissues with oxygen. Many species live longer than their warmer-water cousins as, for example, the cold temperatures slow metabolic rates. Tropical species of krill (e.g. *Nyctiphanes simplex*) usually live only six to eight months, whereas Antarctic krill (*Euphausia superba*) can live up to ten years and their entire life cycle is modified accordingly. Moreover, for species such as Antarctic krill, sea ice is an important habitat and hence they are vulnerable to changes in its extent. There is thought to be 500 million tonnes of Antarctic krill in the Southern Ocean, and they are a vital food source for whales, seals, ice fish, and penguins. Emperor penguins are also vulnerable. Not only does change in sea ice extent affect the abundance of their food, but it also directly affects survival and breeding success because they breed and raise their offspring almost exclusively on the ice. This highlights the sensitivity of the entire polar food chain to the ambient temperatures.

Recent Examples of Extreme Weather

Extreme weather events such as heat waves, droughts, floods, and storms can cause major damage and disruption to human society, with large financial costs and sometimes loss of life. The reinsurance company Munich Re has documented statistics on natural catastrophes worldwide since 1980.[4] Their database shows a substantial increase over the past 35 years in the annual total for the number of recorded weather-related events such as storms and floods. The database also shows a strong upward trend in the total financial losses caused by natural catastrophes, which is thought to be primarily due to socio-economic and demographic factors, such as population growth, ongoing urbanisation, and the increasing value of exposed assets.

The major impact of extreme weather we have seen in the UK in recent years has come from flooding. Around 5 million properties in England, or one in six, are thought to be at risk of flooding and the cost of damage now runs at an average in excess of £1 billion per year. There has been a series of severe floods affecting different parts of the country. The autumn of 2000 was the wettest in England and Wales since records began in 1766 (the longest instrumental series of this kind in the world). The worst affected areas were Kent and Sussex during October and Shropshire, Worcestershire, and Yorkshire in November. Then in 2007, a record was set for the wettest May–July period. Flooding brought terrible damage across England, Scotland, and Wales. Rainfall records were broken again in winter 2014, with flooding across many parts of southern England.

During the 2007 floods, almost 50,000 households and more than 7,000 businesses were flooded, a number of people sadly lost their lives, and billions of pounds of damage was caused. The episode exposed the vulnerability of critical infrastructure to unusual weather: flooding at a Gloucestershire water treatment works left some 100,000 homes without clean water for more than two weeks and an electricity substation supplying power to half a million people came perilously close to being taken off-line. The substation was saved by a massive operation involving the fire service and military to provide temporary flood defences and subsequently £5.5 million has been spent to install more permanent protection. The floods of 2014 further emphasised infrastructure vulnerability and the substantial costs of repairing damage. The exceptional weather caused part of the main train line connecting Devon and Cornwall to the rest of the country to collapse into the sea and thousands of tonnes of material from the coastal cliffs to fall onto the line. The railway was closed for two months while a mammoth operation costing £35 million was mounted to repair the damage. The overall cost of the disruption to the economy was estimated at £1.2 billion. Investigation of options to build greater resilience of this part of the railway network to extreme weather is ongoing, but it is clear that any intervention will come at very considerable cost.

Around the world, extreme weather conditions are leading to temperature and rainfall records being broken, with ever more serious consequences, as what were once extreme conditions are starting to become normal. Freak weather has always occurred, but studies indicate climate change has increased the risk of certain extreme conditions. The connection with an increased risk of heat waves is evident, but analysis also indicates that the kind of heavy downpours responsible for some of the recent flooding in the UK have become more likely because of climate change. In part, this is because a warmer atmosphere holds more water, giving rise in places to more intense rains and increased flood risk.

Sea level rise associated with climate change is also increasing the risk of damage arising from storm surges. Hurricane Sandy hit the Eastern Seaboard of the United States in 2012. New York City was particularly affected, with an estimated $20 billion of damage to the city's building stock and infrastructure. The associated storm surge caused flooding exceeding eight feet above ground level in some locations, with many of the city's subway tunnels being inundated and subway equipment being irreparably damaged by saltwater corrosion. In 2019, the L train line of the New York subway, which handles about 400,000 passenger trips each weekday, will shut for more than a year to repair the damage. The majority of large cities around the world are located in low-lying coastal regions, often for good historical reasons because they were important trading ports. Moreover, many of the developing megacities in Asia and elsewhere are located on or near the coast. Coastal cities are home to more than 10 per cent of the global population, and the numbers are growing rapidly. Thus, many hundreds of millions of people are at ever-increasing risk of coastal flooding due to sea level rise. London is protected from flooding by the Thames Barrier, but for many other cities it would be very difficult or simply too costly to provide a similar defence system.

What Is Leading to Our Changing Climate?

We are seeing climate-related changes occurring around the world and that naturally leads us to question why those changes are happening and what role human activities have played.

There has been a sixfold increase in global population over the past century and a half (Figure 2.1(c)). Back in 1850, the global population was around 1.25 billion. Today, both China and India have country populations greater than this. At the same time, as the population has increased, prosperity has increased enormously. There has been a hundredfold increase in real terms in global GDP (Figure 2.1(d)). While such economic measures may not be a particularly good indicator of societal progress defined in terms of the well-being of people and households, it is nevertheless clear that, in broad terms, significant societal progress has been made over this period in many countries around the world. In terms of the regional distribution of wealth, the United States still dominates the global economy, accounting for nearly a quarter of the world's GDP. The next largest economy is China, but this still lags significantly behind the United States, accounting for about 15 per cent of global GDP according to World Bank data.[5]

Much of the explosion in prosperity since the start of the Industrial Revolution has come about precisely because of that industrialisation, the increase in which can be tracked by increases in energy use. Total global energy use, including all domestic and industrial usage, has increased twentyfold since 1850 (Figure 2.1(e)). This growth was accompanied by a shift from traditional energy sources such as wood, wind, and water power towards fossil fuels, first coal and then oil and natural gas.

In 2016, fossil fuels made up almost 80 per cent of the world's energy use. Hydropower, wood, biofuels made from plants, and nuclear energy together accounted for just under 20 per cent. New renewable energy sources, such as solar and wind, represented about 3 per cent, but their share is growing rapidly. Providing clean, secure, and affordable energy to everyone is one of the greatest challenges of the twenty-first century, as population increase and economic growth cause a rapid rise in demand for energy. More than a billion people worldwide still live without access to electricity, mostly in Africa and Asia. Some 3 billion rely on wood or other solid fuel for cooking or kerosene for lighting, resulting in indoor air pollution that causes millions of deaths each year. Outdoor pollution from burning coal and oil in power plants, industrial facilities, and vehicles causes millions more deaths.

Another striking feature of the past 150 years is the extent to which our activities have changed the land surface at a global scale. One example is the change in land area covered by tropical forest. For thousands of years, humans have cleared forestland to make way for farming. This deforestation was initially most widespread in the temperate regions of our planet. During the last century, deforestation has accelerated in the tropical regions, including the equatorial rainforests, with much of the clearance and degradation being undertaken to make way for industrial farming, e.g. palm oil, soya, and beef. Overall, the area covered by tropical forest has decreased by about 25 per cent since 1850. Cutting down forests disrupts the water cycle, causing worsening drought in some regions, and it is threatening the survival of many animals and plants, including orangutans, gorillas, tigers, and forest rhinoceroses.

These dramatic increases in our global footprint have been associated with a change in the composition of our atmosphere. The atmosphere forms a remarkably thin layer over the Earth; indeed, it is as thin in relative terms as the skin of an apple. It is made up mostly of nitrogen and oxygen but also contains smaller amounts of other gases. These other gases include those commonly referred to as greenhouse gases, which trap the Sun's heat and keep the Earth warm enough for life to flourish, a process known as the greenhouse effect. The atmosphere also contains variable amounts of water vapour, which is a powerful greenhouse gas but which is unique in the sense that the amount of it in the atmosphere is predominantly determined by the temperature. This means that if extra water is added to the atmosphere, it condenses and falls as rain or snow within a week or two. Of the other greenhouse gases, the dominant one is carbon dioxide, but methane, nitrous oxide, and various industrial products are also important. Unlike water vapour, these gases can accumulate in the atmosphere. Because of their heat-trapping influence, increased levels of greenhouse gases in the atmosphere lead to warmer temperatures through an enhancement of the greenhouse effect. Of course, as the temperature changes, the amount of water vapour the atmosphere holds changes, which provides an important feedback. Clouds also influence the passage of heat through the atmosphere and similarly represent an important feedback.

Fossil fuels were formed from prehistoric plants and animals that took carbon from the atmosphere over the course of millions of years. As the fossil energy is burnt, the carbon is being released rapidly back into the atmosphere as carbon dioxide. Forests take carbon dioxide from the air and store it in trees, plants, and soils. When trees are cut down and soil disturbed, much of that carbon is released back into the atmosphere. Indeed, some forests are located on areas of peat soils that contain vast quantities of carbon that accumulated slowly over many thousands of years. When forests are cleared, the soil may be damaged by erosion, and agricultural practices can deplete the soil's organic matter, releasing still more carbon into the atmosphere. Carbon dioxide is exchanged continually between the atmosphere, plants, and animals through photosynthesis, respiration, and decomposition, as well as directly between the atmosphere and oceans. The additional carbon dioxide from fossil fuel burning and deforestation has disturbed the natural balance because the natural restoring processes are too slow compared with the rates at which human activities are adding carbon dioxide to the atmosphere. As a result, a substantial fraction of the carbon dioxide emitted through human activities accumulates in the atmosphere, where some of it will remain for centuries and beyond.

The longest continuous record of atmospheric carbon dioxide levels, dating from 1959, comes from a measurement station at Mauna Loa, Hawaii, and is often called the 'Keeling curve' after the scientist who initiated it and maintained it for many years. Analysis of air trapped in ice sheets and glaciers provides a record of values from earlier times. Together these measurements show that atmospheric carbon dioxide has increased substantially during the modern era (Figure 2.1(f)). In 1750, at the start of the Industrial Revolution, atmospheric carbon dioxide levels were about 280 parts per million (ppm). By June 2016, even the remote Halley Research Station in Antarctica – far from sources of pollution – had recorded carbon dioxide levels surpassing 400 ppm. The global average value of atmospheric carbon dioxide in 2016 was over 403 ppm, about 45 per cent higher than in pre-industrial times. Chemical measurements show there has been a decrease in the fraction of the heavier isotopes of carbon (carbon-13 and -14) and a small decrease in atmospheric oxygen concentration. This indicates that the rise in carbon dioxide is largely from combustion of long-buried fossil fuels, which have low carbon-13 fractions and no carbon-14. Analysis suggests deforestation is responsible for at least a quarter of the increase.

The last time the Earth experienced broadly comparable levels of atmospheric carbon dioxide was during the mid-Pliocene (3–5 million years ago). Today's temperatures are about $1°C$ warmer than in pre-industrial times, but it takes very many centuries for the entire Earth system to fully equilibrate with increased atmospheric carbon dioxide levels. Geological records indicate that during the mid-Pliocene, temperatures equilibrated at about $3°C$ warmer (when referenced to our pre-industrial temperatures). Records also indicate that much of the Greenland and West Antarctic ice sheets melted, and even some of the East Antarctic ice sheet, leading to sea level rise of at least 6 metres and possibly much more. This gives a sense of the scale of long-term change that can arise due to elevated carbon dioxide levels.

Methane is the second-most-important long-lived greenhouse gas, and at present it contributes about one-sixth of the greenhouse effect. The levels of atmospheric methane reached 1,853 parts per billion (ppb) in 2016, an increase of more than 150 per cent since pre-industrial times. The long-term increase is attributed in large part to human activity, including cattle breeding, rice agriculture, landfills, biomass burning, and fossil fuel extraction. After a short period during which atmospheric methane levels appeared to stabilise during the late 1990s and early

2000s, the levels have been increasing again. Evidence from the geographic distribution of changes, and from isotopic measurements, indicates that increased emissions have been strongest from biological sources, most likely associated with tropical agriculture and tropical wetlands, but increased emissions from fossil fuels due to their extraction and use may also play a role.

Through the impact on the greenhouse effect, these changes to the composition of the atmosphere have changed the temperature on Earth. It is worth noting that there is considerably more year-to-year variability in the temperature record than in the records of energy use or atmospheric carbon dioxide. Some of this variability is due to the influence on temperature of factors other than greenhouse gases. Certain volcanic eruptions such as Mt Pinatubo in 1991 can depress global temperatures for several years because of the sulphur aerosols they emit into the atmosphere, and a similar cooling effect can be generated from some air pollutants. Changes in the Sun's activity can produce warming or cooling and 'internal variability', or natural cycles within the climate system, in which heat is transferred between the surface, the air above, and ocean below, that can also produce fluctuations; a well-known example of this is the El Niño phenomenon.

Scientists have developed methods of forensically analysing the data to assess the relative importance of different factors in determining the temperature increase. The pattern of temperature change as it varies geographically and with altitude can be examined, with a different signature or 'fingerprint' of temperature change anticipated according to what the cause is. Increases in the energy received from the Sun result in a general warming everywhere. Conversely, the dust from volcanic eruptions or from some sources of pollution leads to a cooling at the surface and through the lower part of the atmosphere, known as the troposphere, and a warming in the stratosphere above. Increases in greenhouse gases provide a different fingerprint still, with warming at the surface and in the troposphere but a cooling in the stratosphere above. Careful analysis of this nature indicates that the warming over recent decades can be explained through the combined effects of human-made pollution, with some of the warming due to greenhouse gas emissions being offset by the cooling influence of other pollutants, and only a small contribution from natural solar and volcanic influences. This has led to the Intergovernmental Panel on Climate Change (IPCC) concluding in its *Fifth Assessment Report* that 'it is extremely likely that human influence has been the dominant cause of the observed warming since the mid-20th century.'[6]

Putting the recent change into a longer historical perspective provides a sense of how unusual the current changes in climate are. One of the clearest pieces of evidence that tells us about that are the ice cores drilled from Antarctica. As the snow falls in Antarctica, it traps with it air from the atmosphere. As the snow piles up, layer upon layer, this air is trapped as bubbles in the ice. This means that as scientists drill down through the ice, more than 3,000 metres deep, it is like going back in time and they are able to recover the ancient air that was in the atmosphere hundreds of thousands of years ago. It is an amazing museum of the past climate. The air bubbles can be analysed to determine the carbon dioxide levels, and the water in the ice can be analysed to determine the ratio of different isotopes of oxygen, which gives an indication of the temperature in the past.

The longest ice core record we currently have is 800,000 years old, and a European project is currently aiming to drill a core going back 1.5 million years. To put in perspective how far back in time these ice cores go, one of the oldest objects in the British Museum in London is a Stone Age handaxe from the Olduvai Gorge in Tanzania, which is estimated to be 1.2 to 1.4 million years old; Neanderthals are thought to have shared the Earth with us as recently as 40,000 years ago. Hence the ice cores cover the whole of human history and much of pre-history. They also encompass very different climate states. During the last ice age, which peaked about 22,000 years ago, sea levels were some 130 metres below where they are today. By comparison, during the last interglacial warm period, about 125,000 years ago, sea levels were probably somewhere in the range 5 to 10 metres higher than they are today. Over the past 800,000 years, carbon dioxide levels have varied between a low of about 180 ppm during the ice ages to a high of about 280 ppm during the interglacial periods. Today's carbon dioxide levels of over 400 ppm vastly exceed this, clearly demonstrating that the current change lies far outside the natural cycle.

Extreme Weather and Climate Change Now
and in the Future

An annual report issued as a special supplement to the *Bulletin of the American Meteorological Society* aims to explain the extreme weather events of the previous year from a climate perspective.[7] Central to this is asking whether the risk of those weather events occurring has increased as a consequence of the climate change we have seen to date and hence to what extent the extreme weather can be attributed to climate change. One way to address this is to run computer simulations of the climate with and without increased greenhouse gases to assess the difference in the frequency of the extreme weather occurring. This type of 'attribution' research is still in its infancy and the findings often are nuanced and sometimes ambiguous or indeed conflicting, depending on the precise methodology used.

The extent to which people globally are exposed to heat waves depends on the regional patterns both of population and temperature change. The population-weighted average temperature, which accounts for where people live, has been increasing at twice the rate of the global average over recent decades. High temperatures, especially when combined with high humidity, can prove deadly for vulnerable people, especially the elderly. An extreme summer heat wave set temperature records across Europe during June and July 2015. London experienced the warmest July day on record on 1 July 2015, with temperatures hitting 36.7°C (98°F). I remember this well because I was nine months pregnant and became dehydrated; my second daughter was born early the next morning. Subsequent analysis has indicated that the risk of occurrence of this heat wave was enhanced by climate change. Connections between climate change and an increased risk of occurrence have also been found with respect to, for example, damaging heat waves that have struck across China repeatedly in recent years; a severe drought in Ethiopia in 2015 that led to some 10 million people requiring humanitarian assistance; a drought in 2014 affecting parts of Jordan, Lebanon, Palestine, and Israel, which complicated food and water provision for refugees from regional conflicts; and the heat (though not rainfall deficit) associated with an extended hot and dry period in 2011 that is estimated to have cost the Texan economy billions of dollars, especially the agricultural sector.

For the first three-quarters of the past millennium, global average temperatures were reasonably steady. Estimates of the temperatures in the past are based on information from sources such as tree rings, ice cores, corals, marine and lake sediment, and cave formations such as stalagmites, the most knowledge being about the Northern Hemisphere. Compared with the late nineteenth century (AD 1850–1900), there was a period from about AD 950 to 1250 (the Medieval Warm Period) when it is thought that the conditions in the Northern Hemisphere were mostly slightly warmer and a period from about AD 1450 to 1850 (the Little Ice Age) when it is thought the conditions were somewhat cooler. However, the temperature fluctuations were modest, with average temperatures for these two historical periods probably only differing by a few tenths of a degree Celsius (the different reconstructions of temperature detailed in the IPCC's *Fifth Assessment Report* provide different values, with the largest estimate being 0.7°C and the average being 0.3°C). Recent years have been markedly warm in the context of the last millennium. Indeed, the past 30 years, when temperatures have been more than 0.7°C warmer than the late nineteenth century, has very likely been the warmest 30-year period of the past 800 years and probably the warmest of the last 1, 400 years.

Considering the future, Figure 2.2 (left) shows the projections of global average temperature from two sets of computer simulations for the rest of this century. The bands give an indication of the range of values produced by different simulations. The two projections assume different

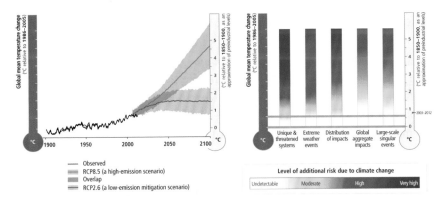

FIGURE 2.2 Future climate change projections (*left*) and risks (*right*)[8]

trajectories for the amount of greenhouses gases in the atmosphere. If we do little to reduce emissions of greenhouse gases, then we are likely to follow a path similar to the 'red' future. This would take us into a world that far exceeds the temperatures of the last millennium. In the 2015 Paris Agreement, the world's governments committed to hold the global temperature to well below 2°C above pre-industrial levels and to pursue efforts to limit the warming to 1.5°C by reducing and eventually eliminating greenhouse gas pollution. The 'blue' future is broadly consistent with this, but it is worth noting that it still puts us at the very top of, or more likely beyond, the temperatures experienced over the past millennium. I think it is worth remembering that these are real people's futures. My daughters will be in their eighties at the end of the century: Figure 2.2 represents their lifetimes. A blue future might not be very different to the life I have lived. A red future would put them in a very different world.

One way to understand what those two futures might hold is to consider the five different categories of 'reasons for concern' that the IPCC collate for their assessment reports, shown in Figure 2.2 (right). Overall, as the global average temperature increases, the risks associated with each of these categories increases.

The first category is 'unique and threatened systems'. Even if we substantially reduce greenhouse gas emissions and future temperature rise is kept modest, in line with a 'blue' future, some ecosystems and cultures will still be at considerable risk. Coral reefs are one example of such a vulnerable system. Corals are severely affected by warming seas – which cause bleaching – and by another consequence of our emissions of carbon dioxide: ocean acidification. About 30 per cent of our emissions of CO_2 each year are taken up by the oceans. That acts as a buffer, meaning that the amount of CO_2 that stays in the atmosphere contributing to warming is less, but the downside is that the oceans are becoming more acidic. Indeed, comparable rates of ocean acidification have not been seen for many millions of years, perhaps not since about 250 million years ago, when the biggest mass extinction of species took place.

The next three categories cover extreme weather and both the distribution and global aggregate of impacts. Climate change-related risks from extreme events, such as heat waves, extreme precipitation, and coastal flooding, are already moderate. A critical challenge is to understand how to build resilience to potentially more extreme conditions in the future. How can cities plan for increased numbers of extremely hot days in terms of their infrastructure, for instance? Providing scientific input to inform such considerations requires ascertaining not only the range of possible future climatic changes at a very local scale, but also the implications of those changes for everything from human health to air-conditioning use and its impact on energy demand. The risks posed by climate change are unevenly distributed both because of the regional variations in projected climatic change and the impacts on, for example, crop yields and water availability and because of variable levels of vulnerability between, and within, different countries. Such a disproportionate burden carries with it all manner of complex moral, ethical, and justice considerations as well as practical challenges. Taking a global view, in many cases the impacts of climate change are expected to increase non-linearly with global average temperature, such that, for example, it is thought that the biodiversity loss associated with a 'four degree world' would be more than twice that associated with a 'two degree world'. An important consideration with respect to all these risks is the correlations between them that can act to accentuate the danger. Extreme weather happening in faraway parts of the world can be connected through global meteorological patterns; storms can be clustered together, or sea level rise can compound flooding from heavy rainfall to exacerbate risks; and climate impacts in different locations can be connected through the global nature of supply chains or societal institutions and infrastructure.

A detailed assessment of the risks posed by climate change to the UK has been undertaken and reported in the national *Climate Change Risk Assessment* that gives a sense of the nature of the threat. The greatest risk in the coming years is thought to come from flooding and coastal change impacting communities, businesses, and infrastructure. This is followed by risks to health, well-being, and productivity from high temperatures and risk of shortages in water supply for public consumption, and for agriculture, energy generation, and industry. Other risks include those to our terrestrial, coastal, marine, and freshwater ecosystems; soils; and biodiversity. Risks to domestic food production are a concern, but perhaps of even greater concern because we live in a globally connected world are the risks to international food trade, with extreme weather affecting supply and prices. Finally, there is a growing unease about the potential for new pests and diseases and invasive non-native species to affect people, plants, and animals.

The final category of reasons for concern is termed 'large-scale singular events' by the IPCC. As the world warms up, there is an increased risk of potentially irreversible environmental changes or abrupt shocks occurring, with global consequences. For example, even modest temperature rise may threaten the vast ice sheets covering Greenland and West Antarctica and lead to seas eventually rising by several metres, transforming coastlines worldwide. There is evidence that during an extended warm period 400,000 years ago, a large fraction of Greenland was ice-free and sea levels rose slowly over centuries to be more than 6 metres higher than today. Local conditions influence ice sheets, but at the time the global average temperature was perhaps only slightly warmer than it is today. Melting has been seen across more than half the Greenland ice sheet during some recent summers, emphasising the current threat. The West Antarctic ice sheet is vulnerable to warm water from the surrounding ocean encroaching beneath it. This is already happening, and there are some indications that glaciers integral to the ice sheet's integrity may now be in irreversible retreat. Other examples of possible abrupt changes about which fears have been raised include the rapid dieback of the Amazon rainforest, mega-droughts, monsoon failures, the collapse of the ocean circulation associated with the Gulf Stream, and the potential massive release of methane – a potent greenhouse gas – from the thawing of vast frozen stores in the Arctic.

What Can Be Done to Limit Future Climate Change?

A very useful simplification can be made to understand the magnitude of future emissions of carbon dioxide that would allow the world to move along a blue pathway of limited temperature increases and associated climate change as opposed to a red pathway. It can be shown that to a good approximation the expected temperature increase is related to the total amount of carbon dioxide we put into the atmosphere over time. This means that the amount of carbon dioxide that can be released before dangerous levels of warming are reached can be seen as a carbon budget.

The entire budget of carbon dioxide emissions for a good chance of staying below 2°C of warming is about 3,000 billion tonnes, but it is estimated that our emissions since 1870 already amount to 2,000 billion tonnes. We have exhausted about two-thirds of the budget already. Emissions today are about 40 billion tonnes per year, and hence at the present rate of fossil fuel use, deforestation, and soil damage we are on course to exhaust the budget for 2°C within the next 20 to 30 years, with the budget for 1.5°C of warming being exhausted even sooner. These numbers have been translated into the proportion of the current proven reserves of coal, oil, and gas that by implication cannot be used given the total is much greater than 1,000 billion tonnes. An estimated value of $20 trillion has been placed on this 'unburnable' carbon, and this has been used to suggest that certain industries are overvalued because they are treating as assets things they could never exploit if we are to stay within a 'two degree' world.

The carbon budget analysis makes clear both the scale and urgency of the global response that is required to address climate change. In 2015, the nations of the world came together in Paris to negotiate an agreement within the United Nations Framework Convention on Climate Change (UNFCCC) dealing with greenhouse gas emissions mitigation, adaptation, and finance.[9] The outcome was unexpectedly bold in terms of the statements that were agreed on. A key aspect was Article 2, which starts by stating,

> *This Agreement, in enhancing the implementation of the Convention, including its objective, aims to strengthen the global response to the threat of climate change, in the context of sustainable development and efforts to eradicate poverty, including by: (a) Holding the increase in the global average temperature to well below 2°C above pre-industrial levels and pursuing efforts to limit the temperature increase to 1.5°C above pre-industrial levels, recognizing that this would significantly reduce the risks and impacts of climate change.*

That 1.5°C statement came as a surprise to many observers, who did not think that the world would agree to such a substantial goal. Equally important was Article 4, which stated that signatories aim to

> *achieve a balance between anthropogenic emissions by sources and removals by sinks of greenhouse gases in the second half of this century, on the basis of equity, and in the context of sustainable development and efforts to eradicate poverty.*

This sets a clear direction to reduce emissions globally to zero in net terms over the course of the next few decades.

The Paris Agreement came into force on 4 November 2016, after enough countries had ratified it. It is the case that the pledges that countries have put forward so far in terms of limiting emissions are not sufficient to meet the Agreement's goals, but there is a process of regular review that it is hoped will lead to countries increasing the level of their ambition.

The UK has domestic targets for reducing greenhouse gas emissions under the Climate Change Act 2008.[10] This established a long-term, legally binding framework to reduce emissions of greenhouse gases (specifically, carbon dioxide, methane, nitrous oxide, and a set of fluorinated gases), committing the UK to reducing emissions by at least 80 per cent below 1990 baselines by 2050, taking emissions from around 800 million tonnes of carbon dioxide equivalent ($MtCO_2e$) per year to around 160 $MtCO_2e$. On the one hand, substantial progress towards this goal has been made (2016 emissions were 468 $MtCO_2e$, just over a 40 per cent reduction), but on the other hand there is still a long way to go to achieve the 80 per cent reduction and then the Paris Agreement objective of net zero emissions.

UK emissions of carbon dioxide have been reduced considerably (over 200 $MtCO_2$) since 1990. In large part this has been due to a decline in the use of coal at power stations and an increase in the use of gas, which has a lower carbon content. Historically, the energy supply has been the largest contributor to UK greenhouse gas emissions. However, substantial reductions in this sector meant that in 2016 the greatest emissions were from the transport sector (126 $MtCO_2e$; 26 per cent share), with the main source of emissions coming from the use of petrol and diesel in road transport.[11] This was followed by energy supply (120 $MtCO_2e$; 25 per cent), business (17 per cent), residential (14 per cent), and agriculture (10 per cent). Clearly, to achieve the further reductions of around 300 $MtCO_2e$ by 2050 will require transformational change across multiple sectors – simply shifting to electric vehicles, for instance, will not in itself be sufficient. Such a requirement for transformational change can be viewed either as a threat or as an opportunity for developing innovative technologies to drive forward low carbon and climate-resilient growth.

Someone who was instrumental in helping to bring clarity to this very complex challenge was Prof Sir David MacKay, a fellow of Darwin College and an inspiration and mentor to many. One of the most powerful aspects of his legacy was his insistence that above all the numbers must add up. His seminal book, *Sustainable Energy – Without the Hot Air*,[12] led to the production of an online 'Global Calculator', a model of the world's energy, land, and food systems to 2050 that allows anyone to explore the world's options for tackling climate change and see how they all add up.[13] A report based on this model found that it is physically possible that a world in 2050 of 10 billion people could eat well, travel more, and live in more comfortable homes, while at the same time reducing emissions to a level consistent with a 50 per cent chance of limiting temperature increase to 2°C.[14] But to do so, we would need to transform the technologies and fuels we use, make smarter use of our limited land resources, and expand forests by some margin. There are significant opportunities for addressing climate change while also providing wider improvements to the quality of people's lives – reducing air pollution, for instance, can have both climate and health benefits. There is great potential for a greener, cleaner, more prosperous future, but it undoubtedly requires strong and visionary technological, social, and political leadership to take us there.

It can seem incredible that our own small actions – the way we heat our homes, the transport we use, our diet, and so forth – are having an impact on global weather systems. And yet the evidence is clear that they are. But the fact that we have been able to exert an influence at a planetary scale in itself shows that we have the ability to reduce and perhaps reverse that influence at a planetary scale. It is surely incumbent on each of us individually and collectively to ensure that happens.

Further Reading

CarbonBrief, *Attributing Extreme Weather to Climate Change*, www.carbonbrief.org/mapped-how-climate-change-affects-extreme-weather-around-the-world (24 July 2018).

CarbonBrief – Clear on Climate, www.carbonbrief.org (24 July 2018).

HRH The Prince of Wales, T. Juniper, and E. Shuckburgh, Annex to *The Ladybird Expert Guide to Climate Change*, www.rmets.org/ladybird-annex/ (24 July 2018).

HRH The Prince of Wales, T. Juniper, and E. Shuckburgh, *Climate Change (A Ladybird Expert Book)* (Penguin, 2017).

D. Liverman, 'Survival into the Future', in E. Shuckburgh (ed.), *Survival* (Cambridge University Press, 2008), pp. 205–224.

J. Slingo, 'Development of climate science', in T. Krude and S.T. Baker (eds.), *Development: Mechanisms of Change* (Cambridge University Press, 2018), pp. 85–107.

B. Watson, 'Risk in the context of (human-induced) climate change', in L. Skinns, M. Scott, and T. Cox (eds.), *Risk* (Cambridge University Press, 2011), pp. 159–180.

Notes and References

1 On the election of President Trump, see Chapter 1, 'Dealing with Extremism', and Chapter 6, 'Extreme Politics: The Four Waves of National Populism in the West' (the editors).

2 Temperature and sea level increases are shown with respect to the 1850–1900 mean. Global GDP is measured in 2010 US dollars; figure by the author.

3 A common British sea spider (*Pycnogonum littorale*) has a leg span of around 20 mm, whereas Antarctic species may have leg spans of up to 750 mm.

4 Münchener Rückversicherungs-Gesellschaft (Munich Re), NatCatService, http://natcatservice.munichre.com/ (accessed 24 July 2018).

5 The World Bank, *World Development Indicators, Gross Domestic Product 2017*, http://databank.worldbank.org/data/download/GDP.pdf (accessed 24 July 2018).

6 Intergovernmental Panel on Climate Change, *Fifth Assessment Report (AR5)*, www.ipcc.ch/report/ar5/ (accessed 24 July 2018).

7 American Meteorological Society, 'Special supplement: Explaining extreme events of 2016 from a climate perspective', *Bulletin of the American Meteorological Society*, vol. 99, issue 1 (Jan., 2018), www.ametsoc.net/eee/2016/2016_bams_eee_low_res.pdf (accessed 24 July 2018).

8 Intergovernmental Panel on Climate Change, *Fifth Assessment Report (AR5)*, Working Group II, *Climate Change 2014: Impacts, Adaptation, and Vulnerability – Summary for Policymakers*, Assessment Box SPM.1, Figure 1, www.ipcc.ch/report/ar5/wg2/summary-for-policymakers (accessed 24 November 2018).

9 United Nations Framework Convention on Climate Change, *The Paris Agreement*, adopted 12 December 2015, https://unfccc.int/files/paris_agreement/application/pdf/parisagreement_publication.pdf (accessed 24 July 2018).

10 United Kingdom, *The Climate Change Act 2008 (c. 27)*, passed into law on 26 November 2008, www.legislation.gov.uk/ukpga/2008/27 (accessed 24 July 2018).

11 For the purposes of the sector-share, calculations of the total (net negative) contribution of emissions and removals of greenhouse gases arising from forest-land, cropland, grassland, settlements, and harvested wood products have been neglected.

12 D. MacKay, *Sustainable Energy – Without the Hot Air* (UIT Cambridge, 2009), www.withouthotair.com/ (accessed 24 July 2018).

13 United Kingdom Department of Energy and Climate Change et al., *The Global Calculator*, http://tool.globalcalculator.org/ (accessed 24 July 2018).

14 United Kingdom Department of Energy and Climate Change et al., *Prosperous Living for the World in 2050: Insights from the Global Calculator* (Crown copyright 2015), www.gov.uk/government/publications/the-global-calculator (accessed 24 July 2018).

3 Probability, Risk, and Extremes

NASSIM NICHOLAS TALEB

This chapter is about classes of statistical distributions that deliver extreme events, and how we should deal with them for statistical inference and decision-making.[1] It draws upon the author's multi-volume series, *Incerto*, and the associated technical research program, which is about how to live in a real world with a structure of uncertainty that is too complicated for us.[2] The *Incerto* tries to connect five different fields related to tail probabilities and extremes: mathematics, philosophy, social science, contract theory, and decision theory, to the real world. The principal idea behind the project is that, while there is a lot of uncertainty and opacity in the world and incompleteness of information and understanding, there is little, if any, uncertainty about what actions should be taken based on such incompleteness, in a given situation.

On the Difference between Thin and Fat Tails

We begin with the notion of fat tails and how it relates to extremes using the two imaginary domains of Mediocristan (thin tails) and Extremistan (fat tails). In Mediocristan, no single observation can really change the statistical properties. In Extremistan, the tails (the rare events) play a disproportionately large role in determining the properties.

Let us randomly select two people in Mediocristan with a (very unlikely) combined height of 4.1 metres – a tail event. According to the Gaussian distribution (and its siblings), the most likely combination of the two heights is 2.05 metres and 2.05 metres. Simply, the probability of exceeding 3 sigmas is 0:00135. The probability of exceeding 6 sigmas, twice as much, is $9:86 \times 10^{-10}$. The probability of two 3-sigma

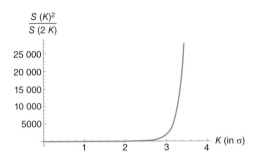

$$\frac{S\,(K)^2}{S\,(2\,K)}$$

FIGURE 3.1 The ratio of two occurrences of K versus one of $2K$ for a Gaussian distribution[3]

events occurring is 1:8 × 10^{-6}. Therefore, the probability of two 3-sigma events occurring is considerably higher than the probability of one single 6-sigma event. This is using a class of distribution that is not fat tailed. Figure 3.1 shows that as we extend the ratio from the probability of two 3-sigma events divided by the probability of a 6-sigma event to the probability of two 4-sigma events divided by the probability of an 8-sigma event, i.e. the further we go into the tail, we see that a large deviation can occur only via a combination (a sum) of a large number of intermediate deviations (the right side of Figure 3.1). In other words, for something bad to happen, it needs to come from a series of very unlikely events, not a single event. This is the logic of Mediocristan.

Let us now move to Extremistan, where a Pareto distribution prevails (among many), and randomly select two people with combined wealth of £36 million. The most likely combination is not £18 million and £18 million. It is approximately £35,999,000 and £1,000. This highlights the crisp distinction between the two domains; for the class of subexponential distributions, ruin is more likely to come from a single extreme event than from a series of bad episodes. This logic underpins classical risk theory as outlined by Lundberg early in the twentieth century and formalised by Cramér, but neglected by economists more recently.[4] This indicates that insurance can only work in Mediocristan; we should never write uncapped insurance contracts if there is a risk of catastrophe. This is the catastrophe principle.[5]

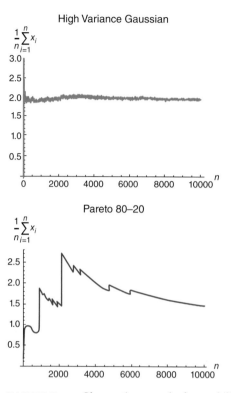

FIGURE 3.2 Observations required to stabilise the sample mean[6]

As mentioned above, with fat-tail distributions, extreme events away from the centre of the distribution play a very large role. Black swans are not more frequent; they are more consequential.

The fattest tail distribution has just one very large extreme deviation rather than many departures from the norm. Figure 3.3 shows that if we take a distribution like the Gaussian and start fattening it, the number of departures away from one standard deviation drops.

The probability of an event staying within one standard deviation of the mean is 68 per cent. As we fatten the tails, to mimic what happens in financial markets, for example, the probability of an event staying within one standard deviation of the mean rises to between 75 and 95 per cent. When we fatten the tails we have higher peaks, smaller shoulders, and higher incidence of very large deviation.

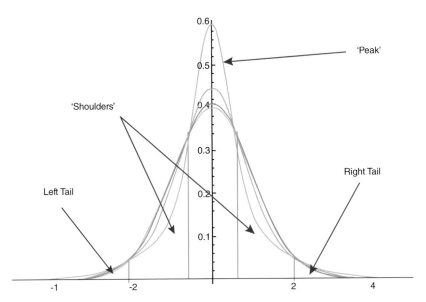

FIGURE 3.3 Fatter tails through perturbation of the scale parameter σ for a Gaussian, made more stochastic (instead of being fixed)[7]

A (more advanced) Categorisation and Its Consequences

Let us now provide a taxonomy of fat tails. Based on mathematical properties, there are three types of fat tails, as shown in Figure 3.4.

First there are entry-level fat tails. This is any distribution with fatter tails than the Gaussian, i.e. with more observations within one sigma and with kurtosis (a function of the fourth central moment) higher than three. Second, there are subexponential distributions satisfying our thought experiment earlier. Unless they enter the class of power laws, these are not really fat tails because they do not have monstrous impacts from rare events. Level three has a variety of names (the power law, slowly varying class, or 'Pareto tails' class) and corresponds to real fat tails.

Working from the bottom left of Figure 3.4, we have the degenerate distribution where there is only one possible outcome, i.e. no randomness and no variation. Above it, there is the Bernoulli distribution, which has two possible outcomes. Then above it there are the two Gaussians. There is the natural Gaussian (with support on minus and plus infinity), and Gaussians that are reached by adding random walks (with compact

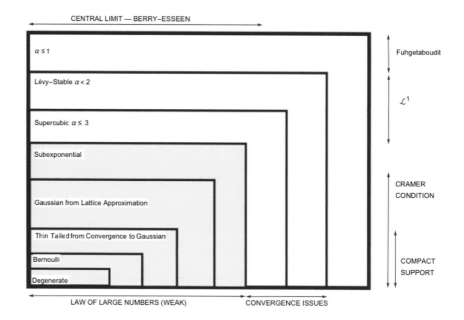

FIGURE 3.4 The tableau of fat tails, along the various classifications for convergence purposes (i.e. convergence to the law of large numbers) and gravity of inferential problems[8]

support, sort of). These are completely different animals, since one can deliver infinity and the other cannot (except asymptotically). Then above the Gaussians there is the subexponential class. Its members all have moments, but the subexponential class includes the lognormal, which is one of the strangest things on Earth because sometimes it cheats and moves up to the top of the diagram. At low variance, it is thin tailed; at high variance, it behaves like the very fat tailed.

Membership of the subexponential class satisfies the Cramér condition of possibility of insurance (losses are more likely to come from many events than a single one), as illustrated in Figure 3.1. More technically, it means that the expectation of the exponential of the random variable exists.[9]

Once we leave the yellow zone, where the law of large numbers largely works, then we encounter convergence problems. Here we have what are called power laws, such as Pareto laws. Then there is one called Super-cubic, and then there is Lévy–Stable. From here there is no variance. Further up, there is no mean. Then there is a distribution right at the top, which I call the *Fuhgetaboudit*. If you see something in that category, you go home, and you do not talk about it. In the category before last, below the top (using the parameter α, which indicates the 'shape' of the tails, for $\alpha < 2$ but not $\alpha < 1$), there is no variance, but there is the mean absolute deviation as an indicator of dispersion. Recall the Cramér condition: it applies up to the second Gaussian, which means we can do insurance.

The traditional statisticians' approach to fat tails has been to assume a different distribution but to keep doing business as usual, using the same metrics, tests, and statements of significance. But this is not how it really works, and these statisticians fall into logical inconsistencies. Once we leave the yellow zone, for which statistical techniques were designed, things no longer work as planned. Here are some consequences of moving out of the yellow zone:

1. The law of large numbers, when it works, works too slowly in the real world (this is more shocking than you think as it cancels most statistical estimators). See Figure 3.2.

2. The mean of the distribution will not correspond to the sample mean. In fact, there is no fat-tailed distribution in which the mean can be properly estimated directly from the sample mean, unless we have orders of magnitude more data than we do (people in finance still do not understand this).

3. Standard deviations and variance are not useable. They fall out of the sample.

4. Beta, Sharpe ratios, and other common financial metrics are uninformative.

5. Robust statistics are not robust at all.

6. The so-called empirical distribution is not empirical (as it misrepresents the expected payoffs in the tails).

7. Linear regression does not work.

8. Maximum likelihood methods work for parameters (good news). We can have plug in estimators in some situations.

9. The gap between disconfirmatory and confirmatory empiricism is wider than in situations covered by common statistics, i.e. the difference between absence of evidence and evidence of absence becomes larger.

10. Principal component analysis is likely to produce false factors.

11. Methods of moments fail to work. Higher moments are uninformative or do not exist.
12. There is no such thing as a typical large deviation: conditional on having a large move, such a move is not defined.
13. The Gini coefficient ceases to be additive. It becomes super-additive. The Gini gives an illusion of large concentrations of wealth. (In other words, inequality in a continent, say, Europe, can be higher.)

Let us illustrate one of the problems of thin-tailed thinking with a real-world example. People quote so-called empirical data to tell us we are foolish to worry about Ebola when only two Americans died of Ebola in 2016. We are told that we should worry more about deaths from diabetes or people getting tangled in their bed sheets. Let us think about it in terms of tails. If we were to read in the newspaper that 2 billion people have died suddenly, it is far more likely that they died of Ebola than smoking, diabetes, or getting tangled in their bed sheets. This is rule number one: 'thou shalt not compare a multiplicative fat-tailed process in Extremistan in the subexponential class to a thin-tailed process that has Chernov bounds from Mediocristan'. This is simply because of the catastrophe principle we saw earlier in Figure 3.1. It is naïve empiricism to compare these processes, to suggest that we worry too much about Ebola and too little about diabetes. In fact, it is the other way around. We worry too much about diabetes and too little about Ebola and other multiplicative effects. This is an error of reasoning that comes from not understanding fat tails – sadly, it is more and more common.

Let us now discuss the law of large numbers, which is the basis of much of statistics. The law of large numbers tells us that as we add observations, the mean becomes more stable, the rate being the square of n. Figure 3.2 shows that it takes many more observations under a fat-tailed distribution (on the right-hand side) for the mean to stabilise. The 'equivalence' is not straightforward.

One of the best-known statistical phenomena is Pareto's 80/20, for example, 20 per cent of Italians owning 80 per cent of the land. Table 3.1 shows that while it takes 30 observations in the Gaussian to stabilise the mean up to a given level, it takes 1,011 observations in the Pareto to bring the sample error down by the same amount (assuming the mean exists).

Table 3.1 *Number of observations required to stabilise the mean*[10]

α	n_α Symmetric	$n_\alpha^{\beta=x\frac{1}{2}}$ Skewed	$n_\alpha^{\beta=\pm1}$ One tailed
1	Fuhgetaboudit	-	-
9/8	6.09×10^{12}	2.8×10^{13}	1.86×10^{14}
5/4	574,634	895,952	1.86×10^{6}
11/8	5,027	6,002	8,632
3/2	567	613	737
13/8	165	171	186
7/4	75	77	79
15/8	44	40	44
2	30	30	30

Despite this being trivial to compute, few people compute it. We cannot make claims about the stability of the sample mean with a fat-tailed distribution. There are other ways to do this, but not from observations on the sample mean.

Epistemology and Inferential Asymmetry

Let us now examine the epistemological consequences. Figure 3.5 illustrates the Masquerade Problem (or Central Asymmetry in Inference).

On the left is a degenerate random variable taking seemingly constant values with a histogram producing a Dirac stick. One cannot rule out nondegeneracy. But the right plot exhibits more than one realisation. Here one can rule out degeneracy. This central asymmetry can be generalised and puts some rigour into statements like 'failure to reject', as the notion of what is rejected needs to be refined.

We have known at least since Sextus Empiricus that we cannot rule out degeneracy, but there are situations in which we can rule out non-degeneracy. If I see a distribution that has no randomness, I cannot say it is not random. That is, we cannot say there are no black swans. Let us now add one observation. I can now see it is random, and I can rule out degeneracy. I can say it is not *not* random. On the right-hand side, we have seen a black swan; therefore, the statement that there are no black swans is wrong. This is the negative empiricism that underpins Western

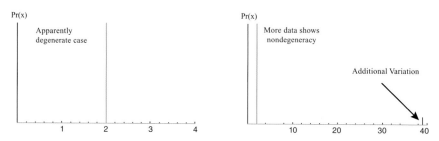

FIGURE 3.5 The Masquerade Problem (or Central Asymmetry in Inference)

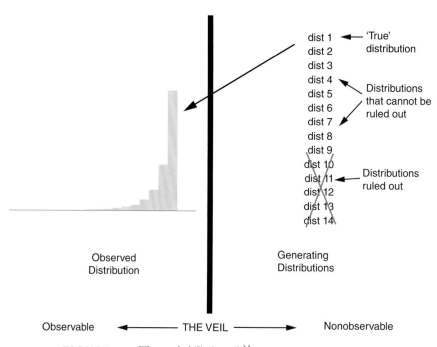

FIGURE 3.6 'The probabilistic veil'[11]

science. As we gather information, we can rule things out. The distribution on the right can hide as the distribution on the left, but the distribution on the left cannot hide as the distribution on the right. This gives us a very easy way to deal with randomness. Figure 3.6 generalises the problem of how we can eliminate distributions.

If we see a 20-sigma event, we can rule out that the distribution is thin tailed. If we see no large deviation, we cannot rule out that it is not fat tailed unless we understand the process very well. This is how we can rank distributions. If we reconsider Figure 3.4, we can start seeing deviation and ruling out progressively from the bottom. These are based on how they can deliver tail events. Ranking distributions becomes very simple because if someone tells you there is a 10-sigma event, it is much more likely that they have the wrong distribution than it is that you really have a 10-sigma event. Likewise, as we saw, fat-tailed distributions do not deliver a lot of deviation from the mean. But once in a while you get a big deviation. So, we can now rule out what is not Mediocristan. We can rule out where we are not. We can rule out Mediocristan. I can say this distribution is fat tailed by elimination. But I cannot certify that it is thin tailed. This is the black swan problem.

Primer on Power Laws

Let us now discuss the intuition behind the Pareto Law. Let x be a random variable. If x is sufficiently large, the probability of exceeding $2x$ divided by the probability of exceeding x is no different from the probability of exceeding $4x$ divided by the probability of exceeding $2x$, and so forth. This property is called 'scalability'.[12]

So if we have a Pareto (or Pareto-style) distribution, the ratio of people with £16 million compared with £8 million is the same as the ratio of people with £2 million and £1 million.

There is a constant inequality. This distribution has no characteristic scale, which makes it very easy to understand. Although this distribution often has no mean and no standard deviation, we still understand it. But because it has no mean, we have to ditch the statistical textbooks and do something more solid, more rigorous.

A Pareto distribution has no higher moments: moments either do not exist or become statistically more and more unstable. So next we move on to a problem with economics and econometrics. In 2009, I took 55 years of data and looked at how much of the kurtosis (a function of the fourth moment) came from the largest observation (see Table 3.3). For a Gaussian, the maximum contribution over the same time span should

Table 3.2 *An example of a power law*

Richer than 1 million	1 in 62.5
Richer than 2 million	1 in 250
Richer than 4 million	1 in 1,000
Richer than 8 million	1 in 4,000
Richer than 16 million	1 in 16,000
Richer than 32 million	1 in 64,000

Table 3.3 *Kurtosis from a single observation for financial data*

Security	Max Q	Years
Silver	94%	46
S&P 500	79%	56
Crude oil	79%	26
Short sterling	75%	17
Heating oil	74%	31
Nikkei	72%	23
FTSE	54%	25
JGBs	48%	24
Eurodollar 1m deposits	31%	19
Sugar	30%	48
Yen	27%	38
Bovespa	27%	16
Eurodollar 3m deposits	25%	28
CT	25%	48
DAX	20%	18

be around 0.008 +/− 0.0028. For the S&P 500 it was about 80 per cent. This tells us that we do not know anything about kurtosis. Its sample error is huge; or it may not exist, so the measurement is heavily sample dependent. If we do not know anything about the fourth moment, we know nothing about the stability of the second moment. This means we are not in a class of distribution that allows us to work with the variance, even if it exists. This is finance.

For silver futures, in 46 years 94 per cent of the kurtosis came from one observation. We cannot use standard statistical methods with financial data. GARCH (a method popular in academia) does not work because we are dealing with squares. The variance of the squares is analogous to the fourth moment. We do not know the variance. But we can work very easily with Pareto distributions. They give us less information, but nevertheless, it is more rigorous if the data are uncapped or if there are any open variables. Table 3.3 debunks all the college textbooks we are currently using.

A lot of econometrics that deals with squares goes out of the window. This explains why economists cannot forecast what is going on. They are using the wrong methods. It will work within the sample, but it will not work outside the sample. If we say that variance (or kurtosis) is infinite, we are not going to observe anything that is infinite within a sample. Principal component analysis (Figure 3.7) is a dimension reduction method for big data and it works beautifully with thin tails.

If there are not enough data, there is an illusion of a structure. As we increase the data (the n variables), the structure becomes flat. In the simulation, the data have absolutely no structure. We have zero correlation on the matrix. For a fat-tailed distribution (the lower section), we need a lot more data for the spurious correlation to wash out, i.e. dimension reduction does not work with fat tails.

Where Are the Hidden Properties?

The following summarises everything I wrote in the *Black Swan*. Distributions can be one tailed (left or right) or two tailed. If the distribution has a fat tail it can be fat-tailed one tail or it can be fat-tailed two tails. And if is fat-tailed one tail, it can be fat-tailed left tail or fat-tailed right tail.

Regarding Figure 3.8: if it is fat tailed and we look at the sample mean, we observe fewer tail events.

The common mistake is to think that we can naïvely derive the mean in the presence of one-tailed distributions. However, there are unseen rare events and with time these will fill in. But by definition, they are low probability events. The trick is to estimate the distribution and *then* derive the mean. This is called plug-in estimation (see Table 3.4). It is not done by observing the sample mean, which is biased with fat-tailed distributions.

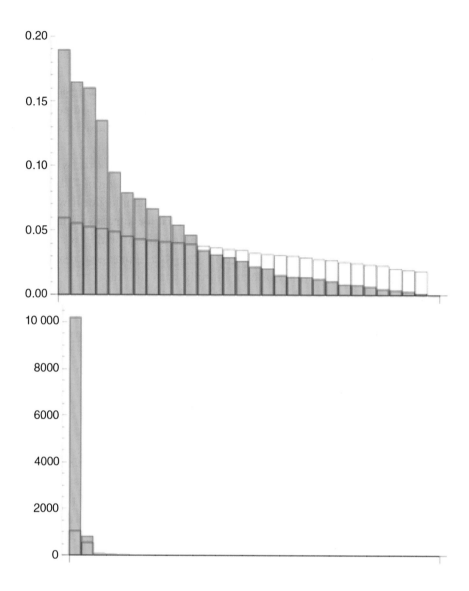

FIGURE 3.7 Monte Carlo simulation showing how spurious correlations and covariances are more acute under fat tails (*bottom*) than Gaussian (*top*).[13] The graph shows principal random principal components under small data sets (solid) and large ones (transparent).

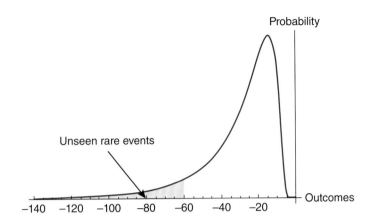

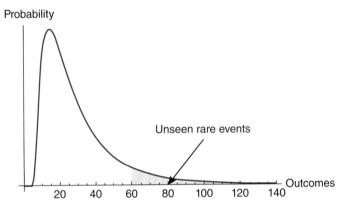

FIGURE 3.8 The biases in perception of risks. Bottom: Inverse Turkey Problem. The unseen rare event is positive. When you look at a positively skewed (antifragile) time series and make inferences about the unseen, you miss the good stuff and underestimate the benefits. Top: The opposite problem. The filled area corresponds to what we do not tend to see in small samples, from insufficiency of points. Interestingly, the shaded area increases with model error.

This is why, outside a crisis, the banks seem to make large profits. Then, once in a while, they lose everything (and more) and have to be bailed out by the taxpayer. The way we handle this is by differentiating the true mean (which I call 'shadow') from the realised mean, as in Table 3.4.

We can also do this for the Gini coefficient to estimate the 'shadow' mean rather than the naïvely observed mean. This is what I mean when I say that the 'empirical' distribution is not 'empirical'.

Table 3.4 *Shadow mean, sample mean, and their ratios for different minimum thresholds (from an example in Cirillo and Taleb's study of violent conflicts)*

Thresh. x 10^3	Shadow mean, x 10^7	Sample mean, x 10^7	Ratio
50	1.9511	1.2753	1.5299
100	2.3709	1.5171	1.5628
145	3.0735	1.771	1.7354
300	3.6766	2.2639	1.6240
500	4.7659	2.8776	1.6561
600	5.5573	3.2034	1.7348

Once we have figured out the distribution, we can estimate the statistical mean. This works much better than observing the sample mean. For a Pareto distribution, for instance, 98 per cent of observations are below the mean. There is a bias in the mean. But once we know we have a Pareto distribution, we should ignore the sample mean and look elsewhere. Note that the field of Extreme Value Theory focuses on tail properties, not the mean or statistical inference.[14]

Ruin and Path Dependence

Let us finish with path dependence and time probability. Our grand-mothers understood fat tails. These are not so scary; we figured out how to survive by making rational decisions based on deep statistical properties.

Path dependence works as follows: if I iron my shirts and then wash them, I get vastly different results compared with when I wash my shirts and then iron them. My first work, *Dynamic Hedging*, was about how traders avoid the 'absorbing barrier', since once you are bust, you can no longer continue: anything that eventually goes bust will lose all past profits.[15] The physicists Ole Peters and Murray Gell-Mann shed new light on this point, and revolutionised decision theory by showing that a key belief since the development of applied probability theory in economics was wrong.[16] They pointed out that all economics textbooks make this mistake, the only exceptions being information theorists such as Kelly and Thorp.

Let us explain ensemble probabilities. Assume that 100 of us, randomly selected, go to a casino and gamble. If the 28th person is ruined, this has no impact on the 29th gambler. So we can compute the casino's return using the law of large numbers by taking the returns of the 100 people who gambled. If we do this two or three times, then we get a good estimate of what the casino's edge is. The problem comes when ensemble probability is applied to us as individuals. It does not work because if one of us goes to the casino and on day 28 is ruined, there is no day 29.

This is why Cramér showed insurance could not work outside what he called the Cramér condition, which excludes possible ruin from single shocks. Likewise, no individual investor will achieve the alpha return on the market because no single investor has infinite pockets (or, as Ole Peters has observed, lives their life across branching parallel universes). We can only get the return on the market under strict conditions.

Time probability and ensemble probability are not the same. This works only if the risk taker has an allocation policy compatible with the Kelly criterion using logs.[17] Peters wrote three papers on time probability (one with Gell-Mann) and showed that a lot of paradoxes disappeared. Let us see how we can work with these, and what is wrong with the literature. If we visibly incur a tiny risk of ruin, but have a frequent exposure, it will go to probability one over time. If we ride a motorcycle, we have a small risk of ruin, but if we ride that motorcycle frequently, then we will reduce our life expectancy. The way to measure this is to focus only on the reduction of life expectancy of the unit, assuming repeated exposure at a certain density or frequency.

Behavioural finance so far makes conclusions from statics, not dynamics; hence, it misses the picture. It applies trade-offs out of context and develops the consensus that people irrationally overestimate tail risk (hence, need to be 'nudged' into taking more of these exposures). But the catastrophic event is an absorbing barrier. No risky exposure can be analysed in isolation. Risks accumulate. If we ride a motorcycle, smoke, fly our own propeller plane, and join the Mafia, these risks add up to a near-certain premature death. Tail risks are not a renewable resource.

Every risk taker who survived understands this. Warren Buffett understands this. Goldman Sachs understands this. They do not want small risks, they want zero risk because that is the difference between the

ENSERBLE PROBABILITY

THE RUIN OF ONE DOES NOT AFFECT THE RUIN OF OTHERS

ROINED

SURVIVED

n OBSERVATIONS

APPROXIMATELY ONE IN n FACE RUIN

TIME PROBABILITY

ONE SPECULATOR ACROSS TIME

RUINED, GAME STOPS

Time

FIGURE 3.9 Ensemble probability versus time probability[18]

firm surviving and not surviving over 20, 30, even 100 years. This attitude to tail risk can explain that Goldman Sachs is 149 years old – it ran as a partnership with unlimited liability for approximately the first 130 years, but was bailed out in 2009, after it became a bank. This is not in the decision theory literature, but we (people with skin in the game) practise it every day. We take a unit, look at how long a life we wish it to have, and see by how much the life expectancy is reduced by repeated exposure.

Next, let us consider layering, or why systemic risks are in a different category from individual, idiosyncratic risks. In Figure 3.10, the worst-case scenario is not that some individual dies. It is worse if your family, friends, and pets die. It is worse if you die and your arch enemy survives. They collectively have more life expectancy lost from a terminal tail event. So, there are layers. The biggest risk is that the entire ecosystem dies (existential risk).

The precautionary principle puts structure around the idea of risk for units expected to survive. Ergodicity in this context means that your analysis for ensemble probability translates into time probability. If it does not, we must ignore ensemble probability altogether.

What Do We Do?

To summarise, we first need to make a distinction between Mediocristan and Extremistan, two separate domains that never overlap with each other. If we do not make that distinction, we do not have any valid analysis. Second, if we do not make the distinction between time probability (path dependent) and ensemble probability (path independent), we do not have a valid analysis.

The next phase of the *Incerto* project is to gain understanding of fragility, robustness, and, eventually, antifragility. Once we know something is fat tailed, we can use heuristics to see how an exposure there reacts to random events: how much a given unit is harmed by them. It is vastly more effective to focus on being insulated from the harm of random events than to try to figure them out in the required detail (as we have seen, the inferential errors under fat tails are huge). So, it is more solid, much wiser, more ethical, and more effective to focus on detection heuristics and policies rather than to fabricate statistical properties.

The beautiful thing we discovered is that everything that is fragile must present a concave exposure similar − if not identical − to the payoff of a short option.[19] There is a negative exposure to volatility. It is

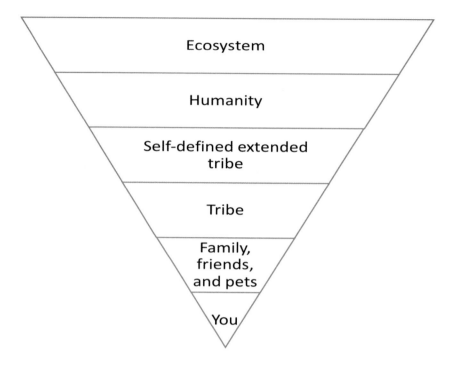

FIGURE 3.10 A hierarchy for survival (tail risk matters more for higher entities that have longer life expectancy)

nonlinear, necessarily. It must have harm that accelerates with intensity, up to the point of breaking. If I fall 10 metres, I am harmed 10 times more than if I fall 1 metre. That is a necessary property of fragility. We just need to look at acceleration in the tails. We have built effective stress-testing heuristics based on such an option-like property.[20]

In the real world, we want simple things that work; we want to impress our accountant and not our peers.[21] My argument in the latest instalment of the *Incerto*, *Skin in the Game*, is that systems judged by peers and not evolution rot from overcomplication. To survive, we need to have clear techniques that map to our procedural intuitions. The new focus is on how to detect and measure convexity and concavity. This is much, much simpler than probability.

Notes, References, and Further Reading

1 The author extends his warmest thanks to Duncan Needham who patiently and accurately transcribed the ideas from this lecture, given at the University of Cambridge on 27 January 2017, into a coherent text. The author is also grateful to Ole Peters, who corrected some mistakes.

2 N.N. Taleb, *Incerto: Antifragile; The Black Swan; Fooled by Randomness; The Bed of Procrustes; Skin in the Game* (Random House and Penguin, 2001–18); for further reading from The Darwin College Lectures, see D. Spiegelhalter, 'Quantifying uncertainty', in L. Skinns, M. Scott, and T. Cox (eds.), *Risk* (Cambridge University Press, 2011), pp. 17–33 (the editors).

3 As K increases, that is, the more we are in the tails, the more likely the event is to come from two independent realisations of K, $P(K)^2$, and the less from a single event of magnitude $2K$.

4 F. Lundberg, *I. Approximerad framställning af sannolikhetsfunktionen. II. Återförsäkring af kollektivrisker*, PhD diss. (Almqvist and Wiksells, 1903); H. Cramér, *On the Mathematical Theory of Risk* (Centraltryckeriet, 1930).

5 For an essay on this principle in applied mathematics from The Darwin College Lectures, see E.C. Zeeman, 'Evolution and catastrophe theory', in J. Bourriau (ed.), *Understanding Catastrophe* (Cambridge University Press, 1992), p. 83 (the editors).

6 Extremistan is represented here by a Pareto distribution with a 1.13 tail exponent, corresponding to the 'Pareto 80–20'.

7 Some parts of the probability distribution gain in density, others lose. Intermediate events are less likely, tails events and moderate deviations are more likely. We can spot the crossovers a1 through a4. The 'tails' proper start at a4 on the right and a1 on the left.

8 Power Laws are in white, the rest in yellow; P. Embrechts, *Modelling Extremal Events: For Insurance and Finance* (Springer, 1997).

9 Let x be a random variable. The Cramér condition for all $r > o$ is $\mathbb{E}(e^{rX}) < +\infty$.

10 The Gaussian case is $\alpha = 2$. The 80/20 requires 1,011 more data than the Gaussian.

11 Taleb and Pilpel explain this from an epistemological standpoint with the 'veil' thought experiment. An observer is supplied with data (generated by someone with 'perfect statistical information', that is, producing it from a generator of time series). The observer, not knowing the generating process, and basing their information on data and data only, would have to come up with an estimate of the statistical properties (probabilities, mean, variance, value-at-risk, etc.). Clearly, the observer having incomplete information about the generator, and no reliable theory about what the data correspond to, will always make mistakes, but these mistakes have a certain pattern. This is the central problem of risk

management, N.N. Taleb and A. Pilpel, 'I problemi epistemologici del risk management', in D. Pace (ed.), *Economia del rischio. Antologia di scritti su rischio e decisione economica* (Giuffrè, 2004).

12 More formally: let X be a random variable belonging to the class of distributions with a 'power law' right tail: $\mathbb{P}(X > x) \sim L(x).x^{-\alpha}$, where $L : [x_{\min}, +\infty) \to (0, +\infty)$ is a slowly varying function, defined as $\lim_{x \to +\infty} \frac{L(kx)}{L(x)} = 1$ for any $k > 0$. We can apply the same to the negative domain.

13 Principal components ranked by variance for 30 Gaussian uncorrelated variables, $n = 100$ and 1,000 data points (above), and principal components ranked by variance for 30 Stable Distributed (with tail = 32, symmetry = 1, centrality = 0, scale = 1) (below). Both are 'uncorrelated' identically distributed variables, $n = 100$ and 1,000 data points.

14 *Supra* note 8 as well as L. de Haan and A.F. Ferreira, *Extreme Value Theory: An Introduction* (Springer, 2006). Springer Series in Operations Research and Financial Engineering.

15 N.N. Taleb, *Dynamic Hedging: Managing Vanilla and Exotic Options* (John Wiley & Sons, 1997).

16 O. Peters and M. Gell-Mann, 'Evaluating gambles using dynamics', *Chaos*, vol. 26, issue 2 (2016).

17 J.L. Kelly, 'A new interpretation of information rate', *IRE Transactions on Information Theory*, vol. 2, issue 3 (Sep., 1956), pp. 185–89; E.O. Thorp, 'Optimal gambling systems for favorable games', *Revue de l'Institut International de Statistique*, vol. 37, issue 3 (1969), pp. 273–93.

18 The treatment by option traders is done via the absorbing barrier. I have traditionally treated this in *Dynamic Hedging* and *Antifragile* as the conflation between X (a random variable) and $f(X)$ (a function of said random variable), which may include an absorbing state.

19 N.N. Taleb and R. Douady, 'Mathematical definition, mapping, and detection of (anti) fragility', *Quantitative Finance*, vol. 13, issue 11 (2013), pp. 1677–89.

20 N.N. Taleb, E. Canetti, T. Kinda, E. Loukoianova, and C. Schmieder, 'A new heuristic measure of fragility and tail risks: Application to stress testing', *International Monetary Fund Working Paper*, vol. 12, issue 216 (Aug., 2012).

21 G. Gigerenzer and P.M. Todd, *Simple Heuristics That Make Us Smart* (New York, 1999).

4 Extreme Rowing

ROZ SAVAGE

This essay is on Extreme Rowing.[1] If you thought it was about extreme rowing (as in arguments), I am sorry, but you are in the wrong place. You will have to tune in to the Brexit negotiations, or the American political scene, for that.[2]

Before I rowed my first ocean, I suppose I thought it would bear some resemblance to the crew rowing I had done at Oxford – but about 3,000 miles further. I was very wrong, for two main reasons. First, even in a good year, that would not be true. Being able to row is only about 1 per cent of what it takes to row across an ocean.[3] The rest is seamanship, survival, logistics, and sheer bloody-mindedness. Second, the year I chose was not a good year. It was 2005, and there were far and away more storms in the Atlantic that year than in any other year since records began – 28 named storms, including Hurricane Katrina, compared with a mere 20 in the next worst year.[4] So my timing could have been better – in fact, it couldn't really have been worse.

Amidst all that whinging, the most pertinent phrase was, 'Just keep going, just keep rowing'. And this is what I want to talk about – that many things that appear extreme are, in fact, the accumulation of a vast number of tiny actions, or sometimes inactions, that together produce massive results – be that individually or collectively. We tend to overestimate how much we can get done in one day – as my overoptimistic to-do list invariably bears witness to – and to underestimate how much we can get done over time with the consistent, and persistent, application of effort.

This law – let us call it the Second Law of Ocean Rowing, the First Law being 'don't leave the boat' – applies to almost any realm of human endeavour. I am specifically going to focus on three areas: monumental and often masochistic feats of physical endurance, personal transform-ation, and creating our collective future as a human species. And further, I will contend that the actions that we choose to take consistently over time – the course we set, in effect – are a function of the inner narratives we hold about who we are, what life is for, and how the world works.

I will draw on several concepts from psychology with a soupçon of neuroscience and illustrate them by reference to the story that I know best – my own life. I offer my story not because I think it is unique, but quite the opposite – I believe that my life and adventures illustrate universal aspects of the human condition. Few of us will be so foolish as to row across oceans, but many of us will ask questions about how we can live a meaningful life and what legacy we will leave when we are gone.

I would like to start by inviting you to join me on one particular weekday evening back in the late 1990s. I am in my early 30s, I have come home late after a long day in the office in my job as a management consultant, and I am sitting at the kitchen table in my little house in West London, writing in my journal. I often write in my journal, but this evening is different. I am trying out a thought experiment loosely adapted from Stephen Covey's *The 7 Habits of Highly Effective People.*[5] I am imagining that I am nearing the end of my life, and, looking back over it, evaluating how I have spent my time on this Earth. I am doing this exercise twice over – the first version being the fantasy, the second being the version that I am heading for if I carry on as I am.

As I write the first version, I think of the people whose obituaries have struck a chord with me. I admire the people who seem to have really got out there and lived life to the full. They are not held back by self-limiting beliefs or by caring too much about what other people think. They might succeed, they might fail, but either way they take on board the lessons learned and push onwards. They live densely, richly, colourfully, cour-ageously. They live lives worth living. Inspired by them, my pen flies across the paper as I imagine the life of my dreams.

The second version – the version that I'm heading for if I carry on as I am – is very different. Safe and superficially successful, but not deep or fulfilling. Compared with the technicoloured fantasy version, it seems rather beige. 'How we spend our days is, of course, how we spend our lives,' said Annie Dillard.[6] And in that moment, as I look at these pages in my journal, I realise that the way I am spending my days is not the way I want to spend my life. Something has to change.

At the time, I thought I was just having an early mid-life crisis. All my friends seemed perfectly happy with their well-paid jobs in management consultancy or investment banking. I thought that maybe there was something wrong with me. I seemed to be the odd one out, the only one for whom this lifestyle didn't work. It was a few years later that I found out that I wasn't special at all – that Jung had identified this concept of individuation, a process of psychological integration and transformation whereby the individual defines themselves as distinct from the general, collective psychology.[7] Of course, I didn't realise that was what I was doing – I just thought I was trying to figure out who I was and what I wanted to do with my life, while also realising that I didn't have forever to do it.

That obituary exercise literally changed the course of my life. It took me a while longer to take action, but once I had glimpsed that alternative – and much more exciting – version of my future, it was only a matter of time before my reality started to align with the vision.

Fast forward about 6 years, and I am a long way from West London. I am alone on a 23-foot rowboat, about 12 miles off the coast of La Gomera in the Canaries. It is 6 pm and getting dark, and I am hanging over the side of the boat, being horribly seasick, and wondering what the heck I am doing here. What am I doing here? It is now several years since I quit my job, since when I have dabbled in various short-lived careers, no doubt convincing my parents and my friends that I have totally lost the plot. I have been a photographer, organic baker, and aspiring coffee shop owner. I have travelled around Peru, discovering Inca ruins, writing a book, and having an environmental awakening.

And that environmental awakening has an awful lot to do with why I am here. After seeing the retreating glaciers in Peru, and living with indigenous Andean people, I have become passionately concerned about the way we are treating our planet, and vowed that, no matter how insignificant my contribution may be, I have to find a way to make a difference.

Shortly thereafter I happened to meet, at the Royal Geographical Society, a young man called Dan Byles who had rowed across the Atlantic – with his mother. His mother. I confess, at the time this didn't strike me as a fun thing to do. The only part that appealed was that every evening, at around sunset, they would have a gin and tonic. That part I could cope with. But the idea must have lodged itself somewhere in my subconscious, because it was just a few months later, when I was driving along in my camper van, minding my own business, that the insane idea popped into my head that I would row across oceans, and use that as my campaigning platform to raise environmental awareness through my blogs, books, and talks.

I am spiritual rather than religious, despite – or maybe because of – having two Methodist preachers for parents, but either way this very much felt like a calling or vocation. At the time, I was not yet aware of Joseph Campbell's book, *The Hero with a Thousand Faces*, but now I know this call to adventure is the first step in the hero's journey.[8] And, true to the classic structure, at first, I refused the call.

I think my heart already knew that this was the perfect project, but my head was coming up with all kinds of excuses – pathetic little details, really, like, 'You've never been to sea before', 'You know you don't like exercise', and 'Who the hell do you think you are – Ranulph Fiennes?' So I spent a week trying to talk myself out of it, but eventually my head caught up with what my heart already knew – that this was exactly what I had been looking for, a perfect way to get on track for that fantasy obituary, and to draw people's attention to our environmental challenges. After all, and very much in keeping with the theme of this book, extreme times call for extreme measures.

So now here I am in my shiny new rowboat, setting out to row 3,000 miles across the Atlantic. I am just a matter of hours into the voyage and already having some pretty serious doubts about the wisdom of the idea. However, I do think there is a lot to be said for having enough naïve optimism to get yourself into something, and then too much stubborn pride to get yourself back out of it. So, although I spend that first night terrified, curled up in the fetal position in my cabin, alone on the dark ocean, being buffeted by large waves, dawn found me back at the oars, still hanging on in there.

This might be a good time to quickly outline some basics of ocean rowing. It had taken me 14 months to get ready for the Atlantic – buying a boat and kitting it out, attempting (and mostly failing) to raise sponsorship, training for up to 16 hours a day on the rowing machine, and taking courses in celestial navigation, basic meteorology, marine communications, first aid, and sea survival. I would be alone for the duration of the voyage, with no support boat, so I had to be entirely self-sufficient, carrying enough food to last 100 days, and a watermaker to turn seawater into fresh water by reverse osmosis. I had a satellite phone to communicate with shore and a little palmtop computer (those things we had before smartphones were invented) to write my blog posts. I could recharge my electronics from batteries connected to solar panels. I had a comprehensive first aid kit and toolkit in case of any problems with my body or boat. The boat itself was made of carbon fibre and was 23 feet long by 6 feet wide, with two enclosed cabins – one for sleeping, one for storage. My bathroom, just in case you are wondering, consisted of a bucket and sponge, and a bedpan.

As to what a typical day looked like, I would not want to miss the sunrise – it was one of the visual highlights of the day – so I would wake up with the first glimmerings of the dawn. I would reach out my hand from my sleeping bag and turn on the GPS to find out where I had drifted overnight. While the GPS was triangulating, I would get out of my bunk and grab a rawfood snack bar to munch on while I filled out the logbook. I would record latitude, longitude, miles to go, wind speed and direction, and the charge in the ship's batteries, and write one line about whatever was on my mind – which varied considerably depending on whether I had gone forwards, backwards, or sideways during the night. Then I would take my rowing pad and a fresh seat cover, plus my rowing gloves, and head out of the sleeping cabin to the deck.

I would row four shifts of 3 hours each, with an hour off in between, and a 10-minute break every hour. During the breaks, I would grab a snack, update the logbook, and tend to the beansprouts that I was growing in a pot to supplement my diet with something fresh. I tried to time my dinner break to coincide with the sunset. I was rowing west across the ocean, but rowers face backwards, so I was always facing east – I definitely got a better suntan on my south side. So, in order to watch the sunset I had to take a break from rowing. And the sunsets were often incredible panoramas – glorious colours and an uninterrupted 360-degree view – that would make me exclaim out loud at their beauty.

FIGURE 4.1 Roz Savage rowing from San Francisco to Hawaii, 2008
(photo credit: Phil Uhl)

My favourite part of the day was after my last rowing shift, when I would brush my teeth while gazing up at the sky. On a clear night the stars could be absolutely amazing, so far away from light pollution. I always watched out for Orion rising, and have never seen the Milky Way so clearly as I did at sea. It made a big difference what moonphase it was – a full moon could be so bright I could see moonshadows and once I even saw a moonbow, while on an overcast night or during a new moon I could hardly see my hand in front of my face.

My last task, which was often a real labour of love because all I wanted to do was curl up in my bunk, was to write and upload my blog update. I would tap it out on the palmtop, add a photo, then link it to my satellite phone and try to coax the phone into maintaining a signal for long enough to upload the email through an incredibly slow and expensive dial-up connection. If the signal dropped, I had to start over again. On a bad night, this process could involve quite a lot of swearing. And yes, just in case you are wondering, as gets reported every time there is news coverage of an ocean rowing expedition, most of us do row naked.

My maiden voyage was quite the sufferfest. There is a sailors' saying that life is easier in the storms, and I found that to be true. At least when I was repairing oars or struggling with the sea anchor, I had something all too real to focus on. For me, the more difficult times came when I was left alone with my thoughts, lamenting the early loss of my stereo – for the first month the weather had been too overcast for my solar panels to generate enough electricity to charge anything other than the essentials of my watermaker and satellite phone. And when the Sun came out, my stereo promptly stopped working due to rust. So for the best part of three and a half months I had nothing but my own thoughts to entertain me.

In his famous memoir, *Walden*, Henry David Thoreau writes: 'It is easier to sail many thousand miles through cold and storm and cannibals, in a government ship, with five hundred men and boys to assist one, than it is to explore the private sea, the Atlantic and Pacific Ocean of one's being alone.'[9] So I cannot help but wonder what he would have made of my decision to simultaneously tackle both the metaphorical inner ocean of solitude and the literal ocean of cold and storms.

This kind of extreme solitude is very interesting. Solitary confinement has long been used as one of the most severe forms of punishment, and in my experience, spending prolonged time alone, especially in a situation involving a considerable degree of fear and stress, can take the mind into some dark places. But it also gives the mind the time and space to think things through, and to come up with ways to be okay with those dark places. To develop the strength to sit in the darkness, without trying to run away from it or shine a light into it, but simply to be there in the lonely recesses of your own psyche, is often uncomfortable but ultimately leads to greater self-knowledge and a confidence in one's own ability to think things through from first principles, to look inside rather than seeking second-hand answers outside, to slow down, concentrate, focus, and pay attention. Like Vipassana on steroids, although not much fun at the time, this is a tremendously empowering – and powerful – process. As William Deresiewicz rather counter-intuitively said, 'If you want others to follow, learn to be alone with your thoughts.'[10]

In our hectic twenty-first-century world, when we are bombarded daily with media, noise, information, conversation, and advertising, when multitasking and busyness are idolised, it was a privilege to step off the Earth and spend several months immersed in nature and solitude. I wouldn't have chosen to have the stereo break, but in retrospect, I am so glad that it did.

Also, life can get quite strange when we are no longer 'in relation to' other people. At times I found myself narrating my own story, describing my actions and my emotions in the third person as if I needed an observer, and in the absence of one I had to create one. At other times, I was perfectly happy to just be. Now that I was not being somebody's child, sibling, friend, colleague, or romantic partner, I felt that I could peel away the layers of these different identities in search of whatever lay at the core. And what did I find at the core?

If you are expecting some cosmic revelation, I hate to disappoint you. I found what felt like ... nothing. Which was not a bad thing – it was actually quite liberating. For blissful moments of time, I would be unaware of being Caucasian, female, British, 30-something years old, Oxford graduate, recovering management consultant ... all these other identities that we layer on to ourselves. I was simply a rower, on a very large ocean, along with a lot of creatures that were much better adapted to this environment than I was.

Life got even more interesting when my satellite phone also stopped working, 24 days before the end of the voyage. I hadn't been using the phone a lot – mostly just a quick daily call to my mother to let her know I was okay. As you can imagine, she had not been delighted when her elder daughter announced she was going to row solo across the ocean, especially as we had lost my father just the year before. Oh, and my sister was travelling alone around the world at the same time and had gone incommunicado in Nepal just when my phone broke. So my poor mother had quite something going on.

Selfishly, though, the demise of my phone was transformative for me, in a very positive way. Up until that point, I had been receiving weather forecasts via text. If it was a bad forecast, I would get anxious about the incoming storm. If it was a good forecast, I would get excited about the good mileage I would make. Most of the time, it was wrong anyway, but meanwhile I had been on this emotional rollercoaster of fearful dread or eager anticipation. To paraphrase Mark Twain, I had a lot of worries on the ocean, most of which never happened.

When the phone broke, my mindset transformed. I became much more present in the moment. From then on it would be just me, my little boat, the ocean, and whatever weather Mother Nature chose to dole out to me.

There was a kind of serenity in the lack of information, no choices to be made other than the moment by moment choice to carry on rowing.

Returning to the hero's journey – and by the way, I am not for a moment suggesting that I am a hero – it is part of the hero's destiny to struggle and almost fail, yet to prevail and come back a better person. According to Campbell's analysis, the hero ventures forth from the ordinary world into an extraordinary parallel world of supernatural wonder, where she encounters fabulous forces and vanquishes dragons. Finally, the hero returns from this magical world with new insights, or boons, to share with her fellow humans. Despite the stormy conditions that year, the dragons I faced were mostly myself. Given that there was only one person on the boat, it was astonishing how many near-mutinies there were, as the lazy devil on my one shoulder battled with the rather irritating disciplinarian angel on the other.

Moving on to the boons, I would like to offer three. They may or may not be new to you, but I shed a lot of sweat and tears, and occasionally blood, to figure these out, so I trust that they were worth the trouble. The first one is:

How to Navigate a Course

On the one hand, it is important to set the right course, because errors get compounded over time, so even if you are just half a degree off at the outset, if you stay on that course then after a few thousand miles you will be a long way from where you wanted to be. On the other hand, no matter how you set out, you will have to check frequently to make sure you are still on track. It is said that an aeroplane is off course 99 per cent of the time, but because the autopilot is constantly correcting for head-winds, sidewinds, updrafts, and downdrafts, the plane still ends up where it is supposed to be.

For me, even though I was course-correcting many times a day, I was very much at the mercy of winds and currents. Sometimes I simply could not go the way I wanted to go, and would have to make best efforts, hoping that conditions would be more favourable later. Other days, I would have to choose between making fast progress aligned with the elements or making a lot less progress but in a better direction, which usually involved bashing sideways across the waves – not very comfortable, but it was often necessary, because the absolute priority was to avoid ending up in the situation where my weatherman would say, 'You can't get there from here,' meaning that I just would not have the horsepower to cut across the winds and currents to where I wanted to be.

We all get buffeted by the winds and currents of life and may well feel that we are off course 99 per cent of the time. In my experience, the most important thing was for me to focus on *why* I was doing this, not *how* I was going to do it. I always had the sense that I would finish the crossing, even when I had no idea how I was going to stay safe and sane that long. It is not that I was courageous – I certainly was not, at least not when I set off – but my motivation was so massive I was able to overcome my fears.

Although I realise it may be controversial to suggest this in such an august institution of learning, I sometimes wonder if it was a curse rather than a blessing that I was good at passing exams. I got very good at jumping through other people's hoops, and when I graduated and looked out at a hoop-less future, I was overwhelmed by the range of opportunities that lay before me. I had no idea how to narrow my options, no way to know where to set my course. When I quit my job as a management consultant, I really did not know what I was going to do next. Those few years that I spent dabbling were an essential phase, during which I figured out a lot about what I wanted to be when I grew up.

A very important step was to get rid of my fears – fear of failure, fear of being an outcast, fear of appearing foolish or overambitious or unrealistic, fear of how it might look on my CV. Warren Buffett, who I believe has done quite well for himself, is said to have advised, 'Take a job that you love. You will jump out of bed in the morning. I think you are out of your mind if you keep taking jobs that you don't like because you think it will look good on your resumé. Isn't that a little like saving up sex for your old age?'

Rather more poetically, Howard Thurman is quoted as saying: 'Don't ask yourself what the world needs. Ask yourself what makes you come alive, and go do that, because what the world needs is people who have come alive.'[11] What the world needs is people who have come alive. Indeed. But for me, there was also an element of wanting to serve some bigger purpose – and that became a crucial motivator. There were many,

many days when if it had been all about my quest to find out what I was capable of, that would not have been motivation enough. I needed the environmental mission as well. But at the same time, I didn't want to be entirely a martyr to the environmental cause – I wanted to grow as a person, too. I definitely needed both the inner motivation and the outer motivation – that, for me, was the sweet spot.

And if I needed validation that I was on the right path, life definitely showed up and delivered. I found that fortune does indeed favour the bold – all kinds of amazing people appeared in my life, and continued appearing, to help make my vision become a reality. Although I am completely solo when I am on the ocean, there is no way I could have done what I have done without the help of countless people who have lent a hand, lent a home, taken care of shopping lists, sponsored a mile, sent a supportive email, managed my social media, or otherwise been an indispensable part of this massive undertaking, and I would like to take this opportunity to acknowledge their huge generosity of time, money, and support over the years.

Once I had set a course, my next challenge – and my second boon – was:

How to Keep on Going

Having rowed around 15,000 miles in all, across three oceans, not counting all the little detours and occasional 'backwards progress', I feel well qualified to advise on this. As I have already mentioned, athletic motivation does not come easily to me. Inside this extreme rower, there is a couch potato trying to get out, and sometimes succeeding. So, I had to develop strategies – and many of them – to keep myself on the rowing seat when I would rather be doing anything else. Of course, it helps that there are not that many things to do on a small rowboat, other than row.

The most successful strategy for me was to turn it into a question of identity, or, to put it a different way, to change the story I was telling myself about who I am. Did I want to be the kind of rower who exhibited discipline, determination, and dedication? Or did I want to be the kind of rower who slacked off and lazed around in my bunk? I also found that

I could reframe situations to make them seem more of a blessing, less of a curse. Here is an example.

I was a month or so into the Atlantic crossing and was having a fairly typical day at the oars. Since my stereo had broken I had nothing but my own thoughts for entertainment and had found that mostly they were not very entertaining at all. Researchers tell us that 90 per cent of our thoughts are the same thoughts we had the day before, and when I was on the ocean that might even have been quite a conservative estimate. All in all, I was pretty sick of myself.

This particular day I was in a very negative frame of mind, focusing on how uncomfortable everything was on my tippy little rowboat. I had tendinitis in my shoulders and saltwater sores on my bottom, and I was popping painkillers like Smarties. My sleeping bag was damp and mouldy, everything took ten times as long to do on a boat as it did on dry land, and the ocean had developed an annoying habit of plopping a wave into my dinner just as I was about to eat it. I am not joking – this happened with way above average regularity.

As my thoughts spiralled around this idea of discomfort, I had a flash of insight. In the run-up to the voyage, whenever a journalist or anybody else asked me why I was rowing the Atlantic, as well as the environ-mental mission, I would also say – rather carelessly, as it turns out – that I wanted to get outside my comfort zone. It now occurred to me that getting outside my comfort zone would, by definition, have to be uncom-fortable. So, my enormous discomfort was not a sign of failure – it was, in fact, exactly what I had wished for.

By reframing my situation in this way, I was able to completely flip my attitude, and spent the rest of the day being quite cheerfully miserable about just how uncomfortable I was. Not for a moment to compare my entirely voluntary suffering with being in a concentration camp, but I am constantly inspired by the wise words of Viktor Frankl in his book, *Man's Search for Meaning.* Even in the midst of the horror of Auschwitz, he was able to reframe his situation and to perceive it as a test presented to him by life. He wrote: 'We needed to stop asking about the meaning of life, and instead to think of ourselves as those who were being questioned by life – daily and hourly ... Life ultimately means taking the responsibility to find the right answer to its problems and to fulfil the tasks which it constantly sets for each individual... When a man finds that it is his destiny to suffer, he will have to accept his suffering as his task; his single and unique task.'[12]

In other words, no matter what is happening, it is entirely possible to reframe one's inner narrative to give oneself a sense of agency and resourcefulness rather than helplessness and victimhood. When I look back over my life story, at any point where I changed course and took the road less travelled, I can see that there was a fundamental shift in my inner narrative – either as a precursor to, or as a result of, the pivot point. A lot of our narrative is subconscious, laid down in early childhood when we have no filter, and we become aware of it only when we find ourselves behaving in counterproductive or self-sabotaging ways. The good thing about those moments is that it brings our narrative into consciousness, and once we're conscious of it, we can rewrite it. We can use our free will to overcome our subconscious conditioning.

My thoughts on this were very much influenced by a book I read round about the time of my personal reinvention, called *Conversations with God*, by Neale Donald Walsch, which I think follows on well from Frankl. Walsch wrote,

> Remember that everything you think, say, and do is a reflection of what you've decided about yourself; a statement of *Who You Are*; an act of creation in your deciding who you want to be.[13]

Whether or not you believe in God doesn't matter, because this statement is quite literally true, which brings me to my third, final, and most wide-reaching boon:

The Power of Accumulation

Since reading Walsch, I have become quite fascinated by this idea of how we create our reality through a multitude of tiny actions, and I'd like to look at that from three different perspectives.

First, I will follow on from Walsch with a bit of neuroscience. Any real neuroscientists among the readers will probably groan at this point, expecting another piece of neuro-codswallop to be offered in a vain attempt to appear intellectual. To the best of my knowledge, by learning new things or forming new habits, we form new neural connections in our brains that essentially hardwire the changes, making them our new habitual way of being. When we first try something new, a new neural connection is forming – an axon reaches out from one neuron and shakes hands with a dendrite from a neighbouring neuron. As we do that new thing again and again, the connection gets stronger. New layers of myelin coat the connection to protect it and accelerate the transmission of the signal from one neuron to the next. This thicker myelin sheath helps improve all types of brain-related tasks, like reading, creating memories, playing a musical instrument, or decision-making. The saying goes that neurons that fire together, wire together. We literally change our brains, through conscious effort that over time becomes less conscious and less effortful.

Even an 'aha!' moment – when something suddenly pops into consciousness with enormous clarity – does not come out of nowhere. Rather, it is the result of a steady accumulation of information. Over time, our level of understanding increases until we suddenly 'get' it. Inventors, scientists, researchers, academics – throughout history our leading minds have had to immerse themselves in their subject and develop these neural networks until, after years of obsessive hard work, inspiration strikes.

Secondly, I would like to look at the power of accumulation in relation to sustainability. When I give talks on an environmental theme, I sometimes get people coming up to me afterwards and saying, 'I take your point – but what can I do? I'm just one person. How can I make a difference?'

And my response is always the same – we are already making a difference, every single day. We have had a number of big environmental disasters – the Deepwater Horizon, *Exxon Valdez*, *Amoco Cadiz*, Bhopal, Chernobyl – but most of our challenges, like climate change, plastic pollution, habitat destruction, overfishing, overpopulation, and so on, are the result of trillions of suboptimal decisions being made by 7.3 billion people every day, every week, every month, every year.[14]

Having rowed around a fairly substantial proportion of the Earth's circumference, at very slow speed, my perception is that the Earth is surprisingly small, and it amazes me that it is still able to support so many of us – although the science shows that it is increasingly struggling to do so, and we are living on borrowed time. But just as we are damaging the ecosphere by a huge multitude of tiny cuts, I believe that we can restore it by creating a multitude of opportunities for healing. Yes, we do need policy change to better protect our world, and we are working on it, but meanwhile there are things we can all do, starting right now, to make a difference.

I often refer to the metaphor of my ocean-rowing voyages, where one oarstroke took me just a few feet, but 5,379,292 of them took me most of the way around the world. (No, of course I am joking – I did not actually count them all.) But I realised that every single one of those oarstrokes was required, and I could only take one oarstroke at a time, one after another. For any of us who advocate for change, progress can often feel so slow as to be imperceptible, and the distant shore of the ocean very far out of sight. All we can do is to keep the faith and keep putting one oarstroke in front of another – checking the GPS every so often to make sure we are still on course.

We have to remind ourselves about tipping points, that change is often subtle and invisible until a critical mass of people have formed a new way of thinking, and there is a sudden sea change. Not unlike neurons forging new networks that lead to an insight, seemingly overnight we emerge from ignorance into the light, and we find it extraordinary that we ever tolerated slavery, legalised racism, religious wars, bigotry – or environmentally unsustainable behaviour.

This view may be unfashionable in an era that prefers the promise of instant gratification, but I would like to contend that the most worthwhile endeavours can be accomplished only through hard graft. So I would like to raise a glass to all those unsung moments, the moments of boredom, frustration, mundanity, disillusionment, and struggle, without which most of humanity's achievements would never have happened, and also to point out to anybody about to embark on such an undertaking that, in my experience, the more you have struggled to achieve your goal, the sweeter the sense of achievement when you get there.

When I say that every action counts, no matter how small, I am not saying that we can take a few small actions, and that is enough. I am saying that every action counts. Every time we buy something, or throw something away, or choose how to travel from A to B, we are casting a vote for the kind of future that we want. And some actions punch higher than others.

You may well have heard the story of a young girl who was walking along a beach on which thousands of starfish had washed up during a terrible storm. When she came to each starfish, she would pick it up and then throw it back into the ocean. People were watching her with some amusement. She had been doing this for some time when a man approached her and said, 'Little girl, why are you doing this? Look at this beach! You can't save all these starfish. You can't begin to make a difference!' The girl seemed crushed. But after a few moments, she bent down, picked up another starfish, and hurled it as far as she could into the ocean. Then she looked up at the man and replied, 'Well, I made a difference to that one!'

I would like to add a postscript to this story. The next time the man was on a beach after a storm, he saw a crowd of people, stooping and throwing, stooping and throwing. As he drew closer, he saw a little girl with a megaphone, calling out, 'Thank you, everybody! We've saved all the starfish on this beach. Time to go to the next one!' And they all piled into their cars and dashed off to the next beach.

As well as doing the right thing in our own worlds, we need to be witnessed, we need to organise, we need to convene, we need to mobilise. This is very much on my mind at the moment – I am currently teaching at Yale in the United States, a country where the rulebook is being rewritten on a daily basis by the current administration. We cannot stand idly by while constitutional rights are violated and environmental legislation repealed. Robert Swan reportedly said, 'The greatest threat to our planet is the belief that someone else will save it.' We all have to step up and play our part.

Thirdly and finally, I would like to reinforce the point that our lives are the sum total of our days. It all matters – in a very literal sense, as our thoughts create the very matter of our brains. This ties back to that obituary exercise that I did all those years ago, when I realised that if I carried on living my life as I was, I was not going to end up where I wanted to be. If we live to be 82 years old plus a bit, we have around 30,000 days on this Earth. When you are young, 30,000 days can seem like forever. I am now approaching the 18,000-day mark, and suddenly the 12,000 that remain to me, if I am lucky, do not seem like so many. I'm doing my best to make each one count. And it is never too late to start living mindfully. John F. Kennedy used to tell the story of the great French Marshall Lyautey, who once asked his gardener to plant a tree. The gardener objected that the tree was slow growing and would not reach maturity for 100 years. The Marshall replied, 'In that case, there is no time to lose; plant it this afternoon!'

In conclusion, everything we think, say, and do is a statement of *who we are* not only on an individual level, but also on a collective level. Who are we collectively, as a human species, deciding to be? What will make us proud of our obituary, our legacy? What collective narrative will serve us best as we go about our daily business, consuming, travelling, talking, writing, influencing, charting a course, choosing actions, forming neural connections, creating our selves, creating our future?

As I see it, we live in an era of extremes. We have extreme opportunities, as technology, medicine, art, and exploration open up realms previously undreamed of. But we also have extreme risks, which threaten to overtake us before we can fully exploit the opportunities. Lǎozǐ said, 'Great acts are made up of small deeds.'[15] And likewise, the extreme response that we now need to our extreme challenges will also be made up of small deeds. None of us is too small, too young, or too old to help create the kind of future that we want. What will you do today to make the right kind of difference?

And, so as not to finish on too serious a note, I'd like to share a video of the highlights of my ocean voyages – completing my solo Atlantic voyage and each of the three stages across the Pacific.[16] By the time I got to the Indian Ocean, I was really quite over filming such things. The point is to show how amazing it feels to do something really difficult, and after many, many oarstrokes, and many struggles, to understand how it feels to dream, and do the extreme, and get safely to the other side.

Further Reading

R. Savage, *Rowing the Atlantic* (Simon & Schuster, 2010).
Stop Drifting, Start Rowing: One Woman's Search for Happiness and Meaning Alone on the Pacific (Hay House, 2013).

Notes and References

1 I would like to thank Julius Weitzdörfer for having invited me to participate, Darwin College for having hosted the lecture this essay is derived from, my manager, Miriam Staley, for coordinating, and my partner, Howard, for having made a cross-country dash from a meeting in Swindon to be able to attend.

2 On both of these topics, see Chapter 6, 'Extreme Politics: The Four Waves of National Populism in the West' (the editors).

3 For an essay on non-ocean rowing from The Darwin College Lectures, see K. Grainger, 'Development of an athlete', in T. Krude and S.T. Baker (eds.), *Development: Mechanisms of Change* (Cambridge University Press, 2018), pp. 20–36 (the editors).

4 For an essay on storms from The Darwin College Lectures, see N. Cook, 'Storms and cyclones', in J. Bourriau (ed.), *Understanding Catastrophe* (Cambridge University Press, 1992), pp. 127–44 (the editors).

5 S.R. Covey, *The 7 Habits of Highly Effective People* (Simon & Schuster, 1989).

6 A. Dillard, *The Writing Life* (Harper Perennial, 1990).

7 C. Jung, *Collected Works of C. G. Jung*, vol. 2, *Psychology and Alchemy* (Routledge, 1980).

8 J. Campbell, *The Hero with a Thousand Faces* (Pantheon Books, 1949).

9 H.D. Thoreau, *Walden* (Ticknor and Fields, 1854).

10 W. Deresiewicz, 'Solitude and leadership' [lecture at the United States Military Academy at West Point in October 2009], *The American Scholar* (Spring, 2010), https://theamericanscholar.org/solitude-and-leadership/%23.W2BIdPZFwwo (accessed 31 July 2018).

11 H. Thurman, *The Living Wisdom of Howard Thurman: A Visionary for Our Time* (Xsounds True, 2010).

12 V. Frankl, *… trotzdem Ja zum Leben sagen: Ein Psychologe erlebt das Konzentrationslager* (Verlag für Jugend und Volk, 1946); English translation as V. Frankl, *Man's Search for Meaning* (Rider, 2004).

13 N.D. Walsch, *Conversations with God, Book 1: An Uncommon Dialogue* (Hodder & Stoughton, 1997).

14 On the causes of climate change, see also Chapter 2, 'Extreme Weather' (the editors).

15 Lǎozǐ, *Dàodéjīng* (China, possibly 4th century BC); English translation as Lao Tzu, *Tao Te Ching* (Hackett, 1993).

16 For an insight into what this looked like from my perspective, you can watch my speaking showreel, which, together with a number of other videos, can be found here: https://youtu.be/vV-j4Y1KOfs (accessed 13 December 2018).

5 Extremes of War: Stories of Survival from Syria

LYSE DOUCET

There are images that make you stop. Over time, after seeing too many, they can make you turn away. Images of extreme destruction and despair have become synonymous with the major conflicts of our time. The ravaged cityscapes of Syria recall the razed streets of Stalingrad during the Second World War. These are places of profound loss and longing. They are also the battles of people just to survive. We see them again and again – not just in Syria, but also in Iraq, Yemen, and beyond. It happens so often that we are left to wonder: have these extreme landscapes become a usual state of affairs? Even worse, have we come to accept that, at least in particular places and at certain times, they are a 'normal' consequence of conflict? Such scarred topography seems to emerge in places far from our tidier lives in more stable societies. But actually, they are all too close. So close we can see these images on social media, on our phones. No one can say, 'We did not know.' We choose not to know or choose to focus our attention elsewhere.

There are other disasters of our time. In South Sudan, Somalia, Northern Nigeria, and Yemen, millions are pushed to the brink of famine by punishing human-made conflicts. Extreme weather also takes its toll – the horrific aftermath of earthquakes, floods, and tsunamis. In the wake of the catastrophic Indian Ocean tsunami of 2004, I found myself asking, 'How do people find the strength to carry on, after this?' Everything, and sometimes everyone they held dear, has vanished. All taken away in seconds by a powerful force of nature, magnified by human activities. And violence, too, descends to unspeakable savagery. In our time, the brutality of the so-called Islamic State or Daesh is regarded as the most extreme of all.

FIGURE 5.1 Scars of war, Aleppo, January 2018
(photo credit: author)

We live in the best of times, the worst of times. Never has our world been so connected, so wealthy, and so educated. Never has the architecture of confronting abuses been so strong: criminal tribunals; human rights organisations; humanitarian law; satellite technology; and social media. And yet, ours is a world where states and non-state actors act with impunity. They get away with it.

This is where a journalist who travels to places that are often broken, too often heart-breaking, and occasionally horrific, looks to the nineteenth-century English biologist and naturalist whose name graces this series of essays. Arguably, Charles Darwin's most famous phrase is the 'survival of the fittest', which he acknowledged was not his own but that of the distinguished English philosopher Herbert Spencer. Darwin himself wrote of a 'struggle for existence ... a struggle for life'. His was an epic sweep across millennia of evolution. His subject was nothing less than the very survival of species; his theories are debated and dissected to this day.

FIGURE 5.2 Desperation of women trapped in Yarmouk, February 2014
(photo credit: author)

FIGURE 5.3 Resilience in ruins, Homs, January 2018
(photo credit: author)

In writing this essay, I have approached it with a measure of Darwin's guiding spirit to explore how reporting from extreme environments is, in essence, also about finding out how people survive – how they struggle for a life. So often, when I embark on another assignment, people ask me, 'How do people live there?' How do they survive conflicts including all-out war? How do they live with hunger, even starvation? How do they bear this overwhelming weight of enormous loss? Extreme environments punish everyone – men, women, children – forcing them to do whatever it takes to survive. Too many people, in too many places, do not survive. But time and again, no matter where I have been, I have seen people, including the youngest, find within themselves the courage and strength to carry on. Even in the worst of places we find some light in the darkness, a measure of hope, a bit of humour, even heroic acts. Survival demands it.

Survival, a profound sense of threat, also pushes people to do terrible things. It can push them to resort to unspeakable acts and unthinkable violence. In this chapter, I draw upon stories from my own reporting in Syria, arguably the world's most complex and consequential conflict, a brutal war that has been called the humanitarian test of our time. This chapter makes no claim to be comprehensive or to offer any overarching theories. These are stories and observations from a reporter's notebooks.

To begin, how should we define 'extreme'? The *Cambridge Dictionary* defines an extreme as the furthest point from the centre. But where do we situate the centre in our advanced twenty-first-century society? At the very least, it means, for many of us, home, food, family, health, schools for children, security for all, even happiness. For those who live in extreme conditions, life is but a search, a struggle, to return to what any of us, what all of us, would call the centre. Such has been the search for millions of Syrians.

In 2011, Syria began its spiral into what has been described as the worst human disaster of the twenty-first century. No one expected that peaceful protests for change, inspired by the momentous events known as the Arab Spring, would escalate into civil war, and then further into a myriad of proxy wars drawing in regional and world powers. No one expected the war would last so long, or cost so much. In 2011, Syria was a middle-income Arab country, with decent roads and schools, a country self-sufficient in pharmaceuticals. It was celebrated for its storied rich culture and exquisite cuisine and boasted some of the world's most precious heritage sites. But Syria was also known for its authoritarian regime. It was a country of few political freedoms, ruled for four decades by President Bashar al-Assad, and his father, Hafez al-Assad, before him. It has since become synonymous with suffering on an epic scale. More than half of Syria's pre-war population of 22 million is a refugee outside the country, forcibly displaced inside, or dead.

The United Nations (UN) stopped counting casualties many years ago. They said they could not keep up; they could not ensure their figures were correct and complete. Monitoring groups, who try to keep counting, say more than half a million lives have been lost. In 2018 they are still counting. This essay does not analyse the evolution of Syria's conflict, nor what might have been done to avoid it. It looks at some aspects of what has become, in seven short years, such an extreme environment. For many who have spent lifetimes working in conflict zones, Syria became the worst they had ever seen. 'I have never seen a war so cruel in all my life,' decried Staffan de Mistura, the UN's Syria envoy whom I first met nearly 30 years ago while covering another destructive war in Afghanistan. 'I have never seen such acts of torture, even in the Balkans conflict,' reflected Carla Del Ponte, the former chief prosecutor for criminal tribunals for both the former Yugoslavia and Rwanda. 'There are no good or bad ones – they are all bad,' is how she described government forces and rebels, 'both committing war crimes as serious and incredible as each other.' Del Ponte was describing the situation in Syria's second city, Aleppo, a city so legendary that Shakespeare mentions it three times. It is a city so rich in culture it has its own genre of Aleppan music and its own distinct cuisine and is classified as a World Heritage Site. 'It took four thousand years to build Aleppo,' remarked Special Advisor to the UN Jan Egeland in November 2016 during the last weeks of a long, brutal battle to control the city. 'Hundreds of generations to build it and one generation to tear it down in four years,' he lamented.

The conflict for the ancient city of Aleppo became an existential battle in a victory both sides knew could change the course of the war. On one side were the forces of President Assad and his Russian and Iranian allies,

FIGURE 5.4 Agony, Aleppo, January 2018
(photo credit: author)

armed with the ferocious firepower of warplanes and artillery. On the
other were rebel militias ranging from jihadi fighters linked to Al Qaeda
to radical Islamists and more moderate groups. Their backers ranged
from Arab Gulf states to Turkey and Western powers.

When the battle for Aleppo ended in December 2016, it marked a
turning point. All the major cities were back in Syrian government hands
after years of shifting momentum on battlefields across the country. But
the cost of victory was enormous. Of all the tactics deployed with
devastating effect, the siege was one of the most powerful, the most
extreme of all. By the middle of 2016, hundreds of thousands of Syrians
were living in besieged areas – towns and cities encircled by gunmen who
stopped essential goods including food and medicine from going in and
stopped desperate people from going out. Most sieges across Syria were
enforced by the Syrian government and combined constant aerial and
artillery bombardment in a brutal tactic that became known as 'surrender

or starve'. The rebels also resorted to this medieval tactic, as did the most extremist group of all, the so-called Islamic State.

Repeated requests by aid agencies to provide desperately needed food and medicine were either ignored or rejected. The UN's Humanitarian Envoy Stephen O'Brien reported that the UN made 113 requests for aid convoys in 2015. Only 10 per cent were given permission to go ahead. Of the 4.6 million people living in besieged or what were called 'hard-to-reach' areas, only 620,000 received help. International humanitarian law prohibits the targeting of civilians. It also prohibits the starvation of civilians as a tactic of war.

It is hard to report first-hand on these sieges because, for the most part, journalists can't go in or out either. The only accounts are usually wrenching videos and testimonies filmed and posted on social media by opposition activists. But the Syrian conflict was also, from the beginning, a battle for the truth of what was happening. There are always counter-

FIGURE 5.5 Rubble, Aleppo, January 2018
(photo credit: author)

narratives and accusations from opposing sides. In years of reporting on Syria's uprising, we often stood on the edge of besieged areas, waiting for people to emerge. Sometimes, it was a small number of families permitted to escape as part of UN-brokered humanitarian pauses. Sometimes, it was an entire community, including the fighters, after negotiated deals between the warring sides. Often, the pleas of families saying, 'We're dying a slow death' pushed fighters to agree to what was, in essence, surrender on the government's terms.

The tides of people always included elderly men and women, bent by exhaustion and hunger, in creaky metal wheelchairs or strapped to wooden carts. Children straggled along, dragging bags with them, tearful and traumatised. Aid agencies waited, just beyond the security barriers, hands outstretched with bottles of water and with stretchers for the weakest. Each time, we asked, 'How did you survive; what did you have to eat?' Each time, their replies were the same. 'We had no bread, no bread for nine months,' one woman told me during one heart-wrenching exit. 'No bread for a year,' said another woman, the strains of her ordeal etched in her gaunt face and the hollow-eyed stares of her children. 'What did you eat?' I would ask. All too often, the answer was 'grass'.

It took time to understand this reply. I heard it again and again. And, then, I eventually realised, it was grass – grass and herbs pulled from the ground, boiled with water in a pot, sometimes with a bit of rice and some spice. In February 2014, a 9-year-old girl, Bara'aa, whose eyes darted back and forth with a haunted look, told me, 'We ate cats.' It was as if she still couldn't quite believe it and wanted me, like her, to be shocked by it. I was. Even Bara'aa's own cat was killed by a sniper, then picked up from the street by a passer-by desperate for food.

Only twice in seven years, after much lobbying for permission from the Syrian government, were my team and I allowed to enter one of these besieged areas. Both times it took us into a neighbourhood south of the Syrian capital, Damascus, called Yarmouk – just five miles from the city centre but a world apart. Decades earlier, Yarmouk had been a refuge for Palestinians fleeing Arab–Israeli wars. And, for Syrians, Yarmouk grew into one of the most vibrant neighbourhoods of the capital, a destination for shopping, dining, and late-night coffee. On the eve of Syria's uprising, it was home to about 160,000 people. It was, a Syrian friend told me, 'a

FIGURE 5.6 Children of war, Homs, January 2018
(photo credit: author)

place where everything made you smile'. Then war took its grim toll. When the siege first took hold in 2013, about 20,000 remained, in increasingly desperate conditions. A refuge had become a prison.

In the spring of 2018, Syrian security forces launched an assault to take it back. Yarmouk's politics have long been complicated, even by the standards of Syria's tangled war. The district is seen as the gateway to the capital. On all sides, men with guns did whatever they could to hold on to it. Many rebel factions including armed Palestinian factions controlled the area. The Syrian military circled the area with checkpoints. And the Islamic State fighters had a foothold on streets nearby. It was a microcosm of the forces tearing Syria apart.

The plight of its people is what first pushed Yarmouk into the headlines, at least for a moment, in the spring of 2014. A small corner of the war became a sad, stark symbol of Syrian suffering. That is how the Syrian war unfolded. Every year, this destructive war had a different

FIGURE 5.7 Ghost town, Yarmouk, December 2015
(photo credit: author)

icon – Homs, Madaya, Darayya, Aleppo, Ghouta all became, for a time, bywords of a merciless war.

Our short visits into Yarmouk were a glimpse into what the UN called the 'deepest circle of hell . . . in the horror that is Syria'. Even after years of seeing Syria's shattered streets, the crush of destruction and desperation was overwhelming. Thousands of people pressed against rough-hewn security barriers. Soldiers struggled to contain a crowd anxious to reach a rare UN food distribution point at the end of a dark canyon of rubble surrounded by ghostly ruins. But there were only 60 food parcels to distribute that day. Behind them, inside blackened shells of buildings, we could see crowds shouting and trying to push forward. Thousands more people could not even reach this narrow corridor with its slim hope of help. A shrivelled woman, veiled in black, mistaking us for aid workers, pleaded with us. 'Please, please take us out, we are dying here,' cried 60-year-old Wafiqa, sobbing uncontrollably as she cradled her lined face in

FIGURE 5.8 Shattered streets, Homs, January 2018
(photo credit: author)

gnarled hands. At the exit, we met 13-year-old Kiffah, waiting with his two younger sisters. With their mother heavily pregnant, they had been given permission to leave. The little boy dutifully put on a brave face, telling me, 'Life is fine, normal.' And then, instantly overcome by the searing pain felt by all, he mentioned 'a little hunger', and suddenly burst into tears. 'There was no bread,' he cried out. Then he could speak no more. The most complicated of politics was reduced to the simplest of needs: a battle for daily bread, and the barest of what we would all regard as the 'centre' of a life worth living.

Months later, we reported, again from Yarmouk, a story that should be so ordinary. Ninth-graders were studying to sit a major set of exams. Like everything else in Yarmouk, this was a Herculean task. It took months of negotiations between the warring sides to allow 120 young boys and girls to leave their families in Yarmouk, board special UN buses, and go to schools in Damascus to sit the Syrian national exams.

FIGURE *5.9* Shells of buildings, Homs, January 2018
(photo credit: author)

FIGURE *5.10* Life without bread: Kiffah, 13 years old, Yarmouk,
February 2018
(photo credit: Robin Barnwell)

FIGURE 5.11 Pain, Yarmouk, February 2014
(photo credit: Philip Goodwin)

FIGURE 5.12 Battling to learn, Damascus, May 2014
(photo credit: author)

If students did not show up, they lost a year of school. As the teenagers emerged from the warren of ruins, running towards waiting buses, there was a rattle of gunfire. There was Mohammad with his gaunt frame and piercing green eyes. 'I have to write my exams,' he insisted. 'I don't want to lose another year.' Mohammad had already lost so much. One day he reached his home to find a rocket had struck. His mother and two sisters were dead. Two other sisters were injured. Next to him stood Khaled, a 15-year-old boy with the confident pose of a cocky street fighter, and the gaze of a vulnerable child. 'Do you ever get scared, Khaled?' I asked. 'We walk to school every day, when there is school, with our hand on our heart because so many people have died,' he replied. 'I am scared. But if we lose our education, we lose our future and then no one can help us.'

The boys told me they studied by sunlight by day, candles by night. There's no electricity in Yarmouk. Khaled told me it's harder still because 'you need food to make your brain work'. What did they eat? Mostly grass. Grass with water, when they could find water. A bit of rice and spices. They spent two weeks in Damascus sitting their exams, and then most went back into Yarmouk, to be with their families. Life, if you can call it a life, goes on.

When I asked Syrian government officials about sieges, they denied there was starvation, insisting that food was getting in, accusing the rebel groups – the terrorists, as they call them – of hoarding food and trapping civilians. When I asked the opposition, they accused the regime of trying to starve them and their families into submission. When I asked the aid agencies, they lamented in private that no one cared about civilians and kept pressing for access. When I asked the people, they were often too afraid to tell the truth about their situation – whether they were living in a besieged rebel-held area or in a government-run shelter after they had escaped or been forced to leave. Often, as a journalist, it was even hard to ask.

In Aleppo in December 2016, during the last weeks of battle there, the pleas from many capitals for a ceasefire, a humanitarian pause, were ignored. But as rebel forces retreated from one district after another, and government forces held fire, thousands of people fled when they could, often in the dark and under fire.

In a derelict industrial zone turned into a displacement centre, we met people who had fled the besieged rebel-held east of the city for the relative safety of government-held West Aleppo. Some of their stories were clear to see: a young boy hobbling on crutches had lost his leg; a teenager in a wheelchair had lost both of them. And we would ask women, children, the elderly – what was life like? How did you survive? Where is your husband, your father? One woman, sitting on a dusty heap of bulging bags, looked at her sister and asked, 'What should I tell them?' What was her story? Was her husband still fighting for the rebels? Was he trying to escape arrest by the government? Or had he been killed, by one side or the other? What was the right answer, in order to survive this ordeal? We started to notice that many women and children just said their husband or father was dead.

War is always ugly and painful. Syria's war has been a story not just of painful sieges but also of horrific massacres. If you enter 'Syria massacres' into your online search engine, you will find that, in most years, there was a massacre of civilians somewhere, every month. Like sieges, it is rare for journalists to witness the immediate aftermath of a massacre. It is not something most people would want to see. It is something you will never forget.

In January 2013, nearly two years into Syria's uprising, we heard reports that more than 100 men, women, and children had been slaughtered in the village of Haswiya in central Syria. At the furthest edge of an idyllic village setting, I pushed open a door into a sprawling family compound and came upon a horrific scene. Charred corpses lay in the yard and were sprawled in the bedroom. Even a plastic bottle of fuel was left at the spot where it had been spilled. From the kitchen, a wide trail of blood swept across the floor where bodies had been dragged away. The washing was still on the clothes line. Food was still on the stove. Soldiers who accompanied us blamed it on Islamist rebels. Villagers in the centre of Haswiya were too afraid to speak. Much later, people who had left the area told us in detail that a pro-government militia, armed gangs known as 'Shabiha', were to blame. 'I saw smoke coming from our neighbour's house and I heard screaming,' one woman recalled. 'I entered my neighbour's house. Women and children, everyone was burning.' Another man, searching for his uncle, recounted: 'His wife and daughters said they locked him in a room and burnt him alive.' We asked an expert, a forensic pathologist, to help us understand why people would act with such cruelty. He looked at the images and called it a brazen act of impunity, cold and calculated. Bodies were left in the open, fires were set, even knowing that the flames would alert people to this crime.

Haswiya was just one massacre in what is a still-growing catalogue in Syria of possible war crimes and crimes against humanity. Many brave Syrians have been collecting data, smuggling out evidence for the war crimes tribunals that they hope will happen, someday.

Syria is not only the most brutal conflict I have covered in 30 years. It has also been the most difficult in which to find what we would call facts, to get as close as we can to the truth. This is another of its paradoxes. Syria is arguably the world's first social media war. Never has there been so much information about a conflict. It has also been waged with the newest advances in satellite technology. And yet, this is a conflict mired in misinformation, misunderstanding, and manipulation. And Syria's war is no longer just about Syria. Foreign powers, with their own interests, keep it going. Syria is no longer a horrible war over there; it is our story too in many ways.

Syria's plight has touched hearts in many places, including Britain. British surgeons have gone to hospitals on dangerous front lines. British chefs have cooked food for Syria, former students who studied Arabic in Damascus held suppers for Syria, aid workers crossed borders and paid with their lives. Good crosses borders. So does bad. People from many countries have also gone to join the ranks of the so-called Islamic State in Syria and Iraq, or have plotted attacks in other cities including London and Manchester. And people, not dangerous but desperate, have also crossed borders, forming part of what has been called the worst refugee crisis since the end of the Second World War. Some are fleeing for their lives, some are looking for better lives. And Syrians make up the biggest numbers by far.

Syria's refugee crisis became Europe's political crisis. For years, every country has been making its own decision on how to respond to this huge exodus. Some, like Germany, have generously taken in more than a million refugees. Countries such as Britain accepted far fewer. My own country, Canada, with its large land mass and a tradition of welcoming immigrants, took in more than 30,000 Syrians in a space of a few months after a new government came to power in late 2015.

FIGURE 5.13 A second chance, Toronto, September 2016
(photo credit: Rachel Price)

To end this essay, I shall share a more hopeful story of a Syrian family that was given the gift of a second chance at a life. It is about a family I met many times in Syria as we filmed a BBC Two documentary in 2014, 'Children of Syria'. The Sabbaghs' family home on the outskirts of Damascus was gutted in the fighting. When we met, they had taken refuge in a dingy storeroom above a shop in the Old City of Damascus. Teenaged Daed was haunted by her nightmares, of seeing beheaded friends visit her in her dreams. It was a life and a future a little girl feared. By 2015, the family had fled to neighbouring Lebanon and we lost contact – until, completely by accident, our paths crossed in a pretty park in the Canadian city Toronto. It was a tearful reunion – of joy and relief.

The family is embracing a new life with all its new opportunities. But this is not a 'happily-ever-after' story like the fairy-tale princesses Daed now draws in her notebooks. She and her three brothers still have nightmares, but they are disappearing with time. Their father is still deeply scarred, still imprisoned by their traumas in Syria including the murder of an infant child. Some Syrian children bounce back with all the rituals of a normal childhood. Others do not. Doctors speak of a

each other's positions along Syria's border. In 2018, the talk is of new wars within Syria's war.

Syria is now so fractured that many Syrians fear their country, as they knew it, is gone – an enormous price to pay for what began as a peaceful uprising for change but ended in such an extreme unravelling. For so many, so little of the Syria they knew survives. Whatever does, they hold fast.

Further Reading

R. Abouzeid, *No Turning Back: Life, Loss and Hope in Wartime Syria* (One World, 2018).

K. Adie, 'Observing conflict', in M. Jones and A. Fabian (eds.), *Conflict* (Cambridge University Press, 2006), pp. 106–24.

L. Anderson, 'Conflict in the Middle East', in M. Jones and A. Fabian (eds.), *Conflict* (Cambridge University Press, 2006), pp. 82–105.

M. de Rond, 'Life in conflict: Soldier, surgeon, photographer, fly', in W. Brown and A. Fabian (eds.), *Life* (Cambridge University Press, 2014), pp. 84–94.

F. Ledwidge, 'Losing the military intervention game', in D. Blagden and M. de Rond (eds.), *Games: The Spectrum of Conflict, Competition, and Cooperation* (Cambridge University Press, 2018), pp. 80–100.

C. Phillips, *The Battle for Syria: International Rivalry in the New Middle East* (Yale University Press, 2018).

6 Extreme Politics: The Four Waves of National Populism in the West

MATTHEW GOODWIN

The rise of national populism has been one of the most striking developments in modern politics. It is partly reflected in the 2016 election of Donald Trump as the 45th president of the United States and the vote for Brexit but mainly in the rise of an assortment of national populists in Europe – Marine Le Pen in France, Geert Wilders in the Netherlands, Nigel Farage in Britain, Viktor Orbán in Hungary, Matteo Salvini in Italy, Alice Weidel in Germany, and Jimmie Åkesson in Sweden, among others. These figures and their movements are not identical, however. While they have much in common on sociocultural issues, such as strong opposition to immigration, Islam, and social liberalism, when it comes to economics they often pursue different lines. Some like Marine Le Pen attack 'savage globalisation' and call for state intervention, while others like Trump or Nigel Farage appear at ease with capitalism and the free market. In Europe, there are also important differences between East and West. While some national populists like Geert Wilders in the Netherlands have sought to cultivate public opposition to Islam by claiming to defend LGBT communities, others like Viktor Orbán in Hungary are hostile to Islam but advocate traditional social conservatism when it comes to the family and rights for same-sex couples. That said, in *National Populism: The Revolt against Liberal Democracy*, Roger Eatwell and I argue that national populists ultimately share an ideological core – they prioritise the culture and interests of the nation and promise to give voice to a people who feel neglected, or are even held in contempt, by distant political elites.[1] In this chapter, I draw upon existing research to set out the four waves of national populism and challenge some popular myths about its support.

Three Waves of Right-Wing Extremism

Three decades have now passed since Klaus von Beyme edited a collection of essays on what he called the 'third wave' of the extreme right in post-war Europe.[2] This was one of the first academic books to explore the emergence of a new generation of parties and is seen by scholars as an important watershed text in the literature. According to von Beyme, the first wave commenced in the aftermath of the Second World War and included an assortment of openly neo-fascist and neo-Nazi groups that typically remained on the fringe or were banned outright by a more 'militant defensive democracy' that had emerged in the post-war era. This was also a period of European political history when social norms against ethnic nationalism and anti-democratic messages were strong. Notable first wave groups included the *Sozialistische Reichspartei* (Socialist Reich Party, SRP) in Germany, which was formed in 1949 but banned three years later by the Federal Constitutional Court. Another was Oswald Mosley's Union Movement, which was formed in 1948 but failed to have an impact (though Mosley's ideas about pan-European organisation would remain influential). Though not mentioned by von Beyme, while the first wave was an electoral failure some of these groups did play an important role as incubators for activists who would later join more successful movements (acting as what social movement scholars call 'abeyance structures', which provide links between different rounds of mobilisation).[3]

The second wave was more populist in nature, and, though slightly more successful, it too failed to engineer a durable revolt. The most notable movement was 'Poujadism' in 1950s France, an anti-establishment populist revolt that attracted significant support from the self-employed and small business owners. Other examples include the Farmers Party in the Netherlands and the National Democratic Party of Germany that was founded in 1964 and had some limited success in Bavaria, Saxony, and Hesse (only narrowly failing to reach the 5 per cent threshold to win representation in the Bundestag in 1969). Also included in the second wave were the Progress parties in Denmark and Norway that emerged in the 1970s while calling for tax cuts and opposing welfare before later turning to immigration. Another was the National Front

(NF) in Britain, which was rooted in an assortment of right-wing and neo-Nazi groups that had coalesced in the 1960s, some of which could trace their lineage directly to Mosley.[4] The National Front peaked in 1979 when it stood more than 300 candidates but never secured a single seat in the House of Commons.

The third wave arrived amid higher unemployment, economic disruption, and the emergence of anti-immigration sentiment in Europe that accompanied the end of the post-war economic boom and rising immigration. It was reflected mainly in the electoral emergence of Jean-Marie Le Pen and the Front National (FN) in France in the mid-1980s, as well as the Freedom Party of Austria that was taken over by Jörg Haider in 1986 and from thereon enjoyed an upsurge of public support. These two movements would be among the most successful in Europe, reflecting the third wave's stronger impact. Notable moments arrived in 1988 when Jean-Marie Le Pen polled over 14 per cent of the vote at a presidential election in France and in 1990 when Haider and his party received nearly 17 per cent at national elections. Both movements would soon reach new heights.

For von Beyme, however, more important than this electoral support were two other developments. First, unlike its predecessors the third wave was anchored in philosophical 'New Right' debates, led by self-described 'metapolitical' writers and thinkers who adopted Gramsci's claim that intellectual hegemony was a precondition for serious power (they often referred to themselves as the 'Gramsci of the right'). Groups like the *Nouvelle Droite* in France and *Nuova Destra* in Italy essentially sought to provide ideas and narratives that could lead right-wing nationalists out of the margins and into the mainstream. Most important was a change of strategy for how to mobilise public opposition to immigration. The 'classic' biological and openly hierarchical forms of racism that had united earlier generations of activists but were also highly discredited in the shadow of the Second World War made way for 'ethno-pluralism', the idea that culturally distinct groups must be kept separate in order to preserve their unique national characters. Described by some as 'differentialism' or 'cultural racism', these arguments about the 'cultural incompatibility' of immigrants, refugees, or settled minorities like Muslims replaced past claims about white racial supremacism. As Jens Rydgren later observed, by adopting ethno-pluralism, and even though its non-hierarchical aspect was often disregarded, national populists could more ably rally anxiety over immigration and ethnic change without being stigmatised as blatant racists.[5]

The second development was increasing internationalisation, which was reflected in stronger intra-movement links at the European level. Writing in the 1980s, von Beyme also linked this trend to an increase of right-wing terrorism and violence that had accompanied events such as the 'Years of Lead' in Italy, a prolonged period of violence and terrorism by neo-fascist and neo-Marxist groups that commenced in the 1960s, involving groups like the New Order, the Mussolini Action Squads, and the Revolutionary Armed Nuclei, the last of which was linked to a major bombing at Bologna's train station in 1980 that killed 85 people and injured more than 200. Notable events would sustain this interest in right-wing violence, including violence against asylum seekers in Germany in the early 1990s, the Oklahoma bombing by Timothy McVeigh in 1995, and the 'nailbomber' David Copeland, who targeted homosexual communities in London in 1999 following his involvement with the British National Party and reading of influential right-wing texts like *The Turner Diaries*. The third wave parties tended to distance themselves from violence and remained far more successful than their predecessors. During the late 1980s and 1990s, parties like the French Front National, Austrian Freedom Party, Northern League in Italy, and Swiss People's Party emerged or enjoyed significant breakthroughs. They would soon be joined by others.

The Fourth Wave

During the first two decades of the twenty-first century, national populists continued to attract rising support. The aforementioned parties were also joined by the likes of the UK Independence Party (UKIP), Sweden Democrats, True Finns, Pim Fortuyn List and the Party for Freedom in the Netherlands, and Law and Justice in Poland, among others. Though many scholars have viewed these as part of the continuing third wave, thereby viewing them in the same light as movements that were active in the 1980s, I argue that since the 2000s we have witnessed the emergence

of a distinct 'fourth wave' – a cycle of mobilisation that can be differenti-
ated from its predecessors on four counts.

First, aside from putting even greater emphasis on the ethno-pluralist
doctrine there have been other important programmatic changes that are
glossed by simply lumping these parties in with the third wave. These
changes also partly reflect a new issue agenda in Europe that, unlike the
1980s, has seen a general rise in Euroscepticism among voters (or at least
the end of the so-called permissive consensus on EU-related issues), as
well as increased concerns over migration, refugees, and religious-based
forms of Islamist terrorism.[6] Against this backdrop, movements in the
fourth wave have put much stronger emphasis on the specific issue of
Islam and the perceived threat from settled and also rapidly growing
Muslim communities and the so-called Islamification of European societies (arguments often linked to the 'Eurabia thesis').[7] This has led some
fourth wave movements to forge alliances with pro-Israel groups and
funders and others to voice strong support for LGBT communities while
simultaneously warning of the 'threat' posed by Islam – a change that
also appears to be having an important effect on their electoral support.[8]
Meanwhile, whereas in earlier decades some national populists saw the
European Union as a useful bulwark against communism, in the early
years of the twenty-first century these parties generally became more
sceptical or opposed outright to the EU and/or further European integration. Indeed, it has now been shown that Euroscepticism is an important driver of modern support for national populism alongside public
concern over ethnic threat and political distrust.[9] Lastly, there have been
other programmatic changes related to economics, although here we find
less consistency across this loose party family. While some parties shifted
to a more protectionist stance, like the Front National in France that has
called for greater state intervention to protect French workers, others
appear more at ease with the free markets while expressing 'welfare
chauvinism', the idea that welfare benefits and rights should be restricted
to the national in-group and kept from threatening outsiders. Either way,
the tendency to view contemporary parties merely as an extended 'third
wave' glosses over these programmatic changes.

Second, the fourth wave more actively distances itself from anti-
democratic norms, including violence. Few parties today allow links to

openly extremist organisations that characterised earlier parties, like the National Democratic Party of Germany (although there are exceptions like Golden Dawn in Greece). Fourth wave parties like UKIP have refused to collaborate with street-based groups like the post-2009 English Defence League and proscribed members of more extreme political parties. Meanwhile, some of the most successful fourth wave leaders like Marine Le Pen ruled out official cooperation with more extreme parties like Golden Dawn, Jobbik in Hungary, or Ataka in Bulgaria. Research has shown that one important reason why fourth wave parties have enjoyed more success than their predecessors is precisely because they have distanced themselves from neo-Nazism, neo-fascism, and crude biological racism.[10]

Third, while distancing from more extreme groups the fourth wave developed closer collaboration at the international level, partly because of their success at elections to the European Parliament that encouraged ideologically similar parties to work together in order to access funding and other resources. In 2007, the short-lived 'Identity, Tradition, Sovereignty' group brought together the French Front National, the Freedom Party of Austria, Greater Romania Party, Flemish Interest, Ataka in Bulgaria, and Italy's Alessandra Mussolini and one member from a small neo-fascist group. In 2015, the more durable 'Europe of Nations and Freedom' was created, which brought together the French Front National, the Austrian Freedom Party, Geert Wilders' Party for Freedom in the Netherlands, the Italian Northern League, Flemish Interest, the Polish Congress of the New Right, and a former UKIP Member of the European Parliament. They would later be joined by a member of the Alternative for Germany (AfD).[11] The group also has links with Tomio Okamura, the Czech-Japanese leader of the Czech Freedom and Direct Democracy movement. Others like the True Finns in Finland and UKIP's Nigel Farage have steered clear of this group, though Farage has close links to the AfD, Sweden Democrats, and Donald Trump (as well as Five Star in Italy, although this is not a national populist movement).

Lastly, the fourth wave has been characterised by even more striking and durable electoral support while the dynamics of this support differ from the past in significant ways. Parties that arose in the third wave, like the French National Front and Austrian Freedom Party, reached new heights and gained more experience of executive power, whether nationally or locally. In 2016, the Austrian Freedom Party's candidate attracted 46 per cent of the vote at the presidential elections, while the next year saw the party finish second at national elections with support from one in four voters and subsequently join a national governing coalition. In the same year in France, Marine Le Pen reached the final round of France's presidential election in which she lost to Emmanuel Macron but still captured nearly 36 per cent of the vote, nearly double what her father had polled in 2002. While national populists have failed to break through in other states – such as Ireland, Portugal, and Spain – over the past decade others have enjoyed record or strong returns in countries such as Denmark, Norway, Greece, Hungary, Poland, Slovakia, and Switzerland. During the first two decades of the twenty-first century there also emerged significant support for national populists in states that used to be considered 'immune' to this trend. During the 1990s, edited volumes that followed von Beyme's routinely included chapters on 'special cases' like Britain, Germany, the Netherlands, and Sweden, where the question to answer was why national populism had failed.[12] Yet each of these democracies have now witnessed major breakthroughs.

In Britain, Nigel Farage and UKIP used local and European Parliament elections to sidestep the first-past-the-post electoral system, winning the 2014 European Parliament elections outright. At the 2015 general election, however, UKIP's nearly 13 per cent of the national vote translated into only one seat in Parliament.[13] Nonetheless, the British case serves as a reminder that national populists do not need elected power to wield considerable influence. UKIP failed to establish a strong parliamentary presence but still played a central role in forcing a referendum on Britain's EU membership and delivering the Brexit vote. Around seven in ten 'Leave' voters had voted for UKIP or considered doing so.[14] Slightly later, in 2017 in Germany, the AfD captured nearly 13 per cent of the national vote and its first (94) seats in the Bundestag. In eastern Germany the AfD, which already held seats in fourteen of Germany's sixteen state parliaments, emerged as the most popular party among men (and the most popular of all in Saxony). The result undermined the claim that national populists could not find success in the country that had produced National Socialism. In the Netherlands, the

Pim Fortuyn List and then Party for Freedom (PVV) have achieved significant parliamentary representation, with the PVV finishing second in 2017 with 13 per cent of the vote, while the Sweden Democrats emerged in 2010 to win their first (20) seats in the Riksdag before more than doubling this number in 2014, when they polled 13 per cent of the vote.

There have also been some important changes to the dynamics of this support. Some like the French Front National have polled more strongly among women, proving able to close the so-called gender gap that has historically characterised their electorates.[15] In broad terms, since von Beyme's writing in the 1980s these movements have become far more adept at mobilising support from manual workers, often becoming the most popular choice for this group (as was evident in states like France in the mid-1990s, or in Britain between 2013 and 2015).[16] While most parties draw their core support from workers, they have also reached into low-skill service-sector workers, clerks, and (as is the case for the most successful parties) parts of the middle class. That said, the core supporters of national populism tend to be working class, occupying a precarious position, and tend to lack degrees.

The Crisis Narrative: Three Misleading Claims

How can we explain their rise? Particularly in recent years many observers have drawn a straight line from the fourth wave of national populism to the post-2008 global financial crisis, Great Recession, sovereign debt crisis in Europe, and subsequent austerity measures that were pursued in many states following pressure from external organisations and financial markets.

This 'crisis narrative' is popular on the left where complex questions about national identity, belonging, immigration, and borders are often reduced to transactional debates about resources, inequality, and redistribution. The argument is not new. The tendency to trace political extremism or radicalism to economic crises has been shaped strongly by interwar fascism and Nazism that arose against the backdrop of the Wall Street Crash in 1929, Great Depression, and rampant unemployment. Yet this is also partly a misreading of history. In Italy, though strikes sparked anxiety over the threat from communism and an uncertain economic future, when Mussolini and the fascists took power the country was not engulfed by a crisis. In Germany, while a severe depression was present, recent research suggests that it was not people who were hit hardest by the crisis who swung behind Hitler and the Nazis. While the unemployed and those on the bottom of the economic ladder tended to support communists and social democrats, it was the working poor who were hurt economically by the crisis, but who were also at little risk of unemployment because they often owned their own businesses. Among the most likely to switch over to the Nazis were self-employed shopkeepers, small farmers, lawyers, and other professionals and domestic employees.[17] Nonetheless, the idea that the fourth wave of national populism has been fuelled by economic crisis has gained momentum. This was also encouraged by events in Greece, where in 2012, against the backdrop of the Eurozone crisis and harsh austerity, public support for the neo-Nazi Golden Dawn movement rocketed from 0.3 per cent in 2009 to 7 per cent in 2012, enough to hand the movement its first (21) seats in parliament. These were not the first fascists to win election in post-war Europe. In 1992, for example, Mussolini's granddaughter won election in Naples and would later serve in the Senate and European Parliament. But the sudden rise of Golden Dawn fuelled a broader belief that the crisis was leading to a revival of fascism.

'The politics of populist anger,' wrote the *New York Times* in 2013, 'are on the march across Europe, fueled by austerity, recession and the inability of mainstream politicians to revive growth.'[18] Writers at *The Guardian* similarly argued that such parties were being 'force-fed by the worst world recession since at least the 1930s, and possibly since before 1914. Mass unemployment and falling living standards in the euro-area and the wider EU made worse by the crazy and self-defeating austerity obsession of European leaders has opened the door to the revival of the far right'.[19]

The crisis narrative has encouraged three further claims about national populism: first, that these revolts are rooted in events that unfolded after 2008; second, that national populists are mainly fuelled by whites who are on low incomes, on welfare, or unemployed; and, third, that these voters are mainly 'old angry white men' who are struggling to adapt to the modern global economy in which university degrees and high skill levels are the new currency. The implication, we are often told, is that the angry white man will soon be replaced by tolerant and more highly educated millennials.

Each of these claims is linked closely to the 'Left Behind thesis', the idea that populist revolts appeal mainly to low-income white voters who are economically marginalised. This is, however, a misinterpretation of the original thesis that focused as much on the role of values and how these are shaped by things like educational attainment.[20] Yet the narrower income-focused interpretation flourished after the Brexit and Trump victories in 2016, which many observers traced to economically impoverished whites. Some pointed to the fact that the average Leave vote among voters on the highest incomes (i.e. earning more than £3,700 each month) was 38 per cent, whereas among those on the lowest incomes (i.e. less than £1,200 each month) it was 66 per cent. Similarly, when two economists sat down to examine what they called the 'Brexit-Trump Syndrome', they focused exclusively on incomes and the crisis, arguing how 'there can be little doubt that in Michigan and Merthyr Tydfil, South Carolina and Sunderland, the dissatisfaction of people on below-average incomes drove the outcome'.[21]

This focus on left-behind, low-income whites was entrenched by a parallel discussion about the 'politics of despair', which linked populism to economically disadvantaged communities where marginalised whites were effectively killing themselves. One year before Trump's victory, one study found that mortality rates among middle-aged white Americans who did not have college degrees had increased significantly, most likely due to alcoholism, drug addiction, and/or suicide. Some drew parallels with the only other documented case of an increase in the mortality rate for a large demographic group in the Western world, which was Russian men who spiralled into alcoholism and/or suicide following the collapse of the Soviet Union. During the primaries in 2016, the *Washington Post* pointed to an 'eerie correlation' in voting data, suggesting that the insurgent Trump was performing strongest in counties where middle-aged whites were dying the fastest. The renegade outsider had tapped into a profound sense of despair among poorly educated whites who had been left behind and cut adrift by economic change.

Linked to this are arguments about generational change, namely, that national populism is fuelled by angry old (white) men. Consequently, its potential will diminish as each year passes. The rise of highly educated millennials whose social networks are more ethnically, religiously, or culturally diverse, and who are often more tolerant, alongside newly ascendant groups like ethnic minorities, will combine to reduce the amount of space for national populists. Those who make these arguments point for instance to the fact that the average Brexit vote among pensioners was over 60 per cent but among 18- to 34-year-olds it was only 30 per cent. Such facts even led one commentator to calculate that if you assume that birth and death rates in Britain remain constant and a cohort effect stays in place (whereby the young will retain their pro-EU views) then on current projections Remain will hold a clear and overwhelming majority in 2021!

These arguments are linked closely to the theory of value change put forward by Ronald Inglehart in the 1970s. Older white men who hold more traditional values and who formed the cultural majority in the West in earlier decades are gradually being replaced by their children and grandchildren who subscribe to more progressive values. The argument is seductive and often cited by outlets such as the *Economist* when claiming that the populist threat will inevitably diminish as each generation replaces the previous one. Consistently, older cohorts are the most likely to object to their relatives marrying a Muslim, to favour reductions in immigration, and to think that being born in Britain or having British ancestors is a very important marker of national identity, while believing attempts to give equal opportunities to same-sex couples have gone too far.

The implication is twofold. First, to counter national populism governments should focus on restoring economic stability, growth, and employment, as well as boosting conditions for white workers on low incomes, on welfare, and in poverty. The second is that countering these revolts becomes a waiting game.

But these broader claims about national populism are deeply misleading. Interrogating each of them reveals why our popular debates are often far too simplistic and the drivers of this phenomenon more complex. Consider the life cycle of national populism. The phenomenon that found its expression in Brexit and Trump emerged long before the collapse of Lehman Brothers. It was in the 1980s and 1990s when the third wave attracted significant support, while in the late 1990s and 2000s parties like the Front National in France and Austrian Freedom Party reached new peaks.

While national populism was a long time coming, it has also endured amid different economic cycles. According to a recent paper that examines its electoral evolution between 1990 and 2008, most of this advance took place between 1990 and 2008 – before the onset of the Great Recession. During the most severe period of the crisis, between 2009 and 2013, national populists gained an average of only 1.2 points, hardly a major boost. Furthermore, they often recorded their strongest results in parts of Europe that had been relatively spared from the worst effects of the crisis. Regions with high unemployment during the crisis saw declining support for the radical right, whereas regions with still relatively low unemployment rates during the crisis saw its vote shares increase, and those regions that maintained high growth rates saw the strongest support of all.[22] The fourth wave was well established long before the financial meltdown and often did not need the crisis to gather support.

Or look at this from another angle. If we compare support for the fourth wave of national populism during the three years before the outbreak of the Great Recession and the three years after the crisis, we do not find much. In fact, if we looked across the entire EU, the first thing to note is that around ten EU member states do not have successful national populists despite the fact that many of the same states were hit hard by the crisis and the events that unfolded, like Cyprus, Ireland, Portugal, and Spain. Many of these same states also recorded some of the sharpest declines in political trust. Even in Greece, the hardest hit country, Golden Dawn has still 'only' recruited support from one in ten voters. Conversely, of the seven states that have seen the most rapid gains for national populists since 2004, many, like Austria and Denmark, remain well below the average rates of unemployment, debt, and reductions in GDP. Indeed, two of the most successful national populist revolts have taken place in Austria and the Netherlands, two states that have had some of the lowest unemployment rates in Europe. In affluent and stable

Switzerland, it was in 2007 – on the eve of the crisis – that the Swiss People's Party attracted nearly 29 per cent of voters, its highest share on record. In Britain, UKIP achieved its first major breakthrough in 2004, capturing 16 per cent of the vote at European Parliament elections after 48 quarters of growth. This is not to say that the crisis or worry about economic loss is not important. In Europe, one recent study suggests that the financial crisis exacerbated the divide between the haves and the have-nots, encouraging an erosion of support for the main parties and higher levels of volatility.[23] But while it may have further enlarged space for national populism, it was clearly not the major cause.

Subjective worries also appear to be just as important as objective economic distress. Support for Brexit was much stronger among those who subjectively felt they had been left behind. The average Leave vote among those who felt they were living comfortably was 41 per cent, but among those who felt they were just getting by it was 60 per cent and among those who said they were finding it quite or very difficult to get by it was 70 per cent. Trump similarly polled well among average income voters and among some voters on above average incomes, but he did especially well among voters who felt that their economic and social position was deteriorating relative to other groups in society.

Nor do these movements merely recruit the unemployed, those on low incomes, and/or people on the very bottom rung of the economic ladder. Rather, over the past 30 years social scientists have often found that skilled and semi-skilled manual workers – i.e. people who are in employment – are the most likely to support these parties. One influential study, for example, estimated that just prior to the eruption of the global financial crisis parties like the Freedom Party of Austria and the Front National in France drew the bulk of their support from a fairly broad alliance of manual workers, low-skilled service-sector workers, clerks, and the self-employed, while sociocultural middle-class professionals in sectors such as media and the arts were noticeably under-represented.[24]

How can we explain this support? While the crisis narrative is undermined by the basic life cycle of national populism and evidence on the dynamics of its support, it also says little about the role of value conflicts in this political mobilisation. Today, there is an emerging consensus in the academic literature that national populism owes as much if not more to perceived conflicts over values than to conflict over economic resources and how these are underpinned by educational divides.

Politics today is no longer underpinned by a neat dividing line between 'left and right', between socialists on one side and traditional conservatives or classical liberals on the other. These older divisions over economic redistribution have been joined, and are now possibly being eclipsed, by new divisions over competing values, between those who share a more universalistic, individualistic, and socially liberal outlook, and those who hold a more traditionalist outlook, who put a premium on order, stability, conformity, and group identities.

One useful starting point is in 1971, when Ronald Inglehart examined how people's value priorities were changing in the West. Inglehart argued that because of economic growth, the welfare state, an expansion of university education, and greater access to information, people were becoming more likely to exhibit post-material values and less likely to adhere to materialist values. This was what he called the 'silent revolution' theory of value change. Writing at a time when there were few successful national populists in the West, and in the shadow of the New Left revolts during the 1960s, Inglehart argued that 'a transformation may be taking place in the political culture of advanced industrial societies. This transformation seems to be altering the basic value priorities of given generations as a result of changing conditions influencing their basic socialization'.[25]

In the years that followed, analyses shed further light on this intergenerational shift from materialist to post-materialist values. Among older groups, materialist values that focused on economic and physical security were predominant, while among younger ones post-materialist values that stressed things like autonomy, human rights, quality of life, and self-expression were increasingly widespread. As economic growth swept across the West, citizens felt more secure and became more concerned about new issues. Unlike older and less well-educated materialists who worried about how to secure scarce resources, fend off competition, and protect their nation state from threatening outsiders, more secure and degree-holding post-materialists turned to issues like environmentalism, multiculturalism, and how to defend and expand rights for women, same-sex couples, refugees, immigrants, and ethnic minorities. This had clear political repercussions. In several democracies, the evolving value priorities facilitated the emergence of new challengers on the left, like Green and left-wing populist parties.

But not everybody participated in the silent revolution. Beginning in the 1970s and 1980s, the start of a backlash to this cultural change became clearly visible and was reflected in growing levels of support for the third wave of national populism, in which a loose alliance of left-behind voters began to rally around calls for a reassertion of more traditionalist values. These voters typically lacked degrees and tended to share more socially conservative or authoritarian values.

These groups often had quite different life experiences, but they shared a rejection of the shift towards a more individualistic society in which the dominant liberal consensus supported if not celebrated things like mass immigration, transnational and global identities, and European integration. In this sense, the opposition to immigration or Islam that dominates the outlook of national populist voters is only one part of a much broader, values-led backlash.

One of the first people to spot this backlash was the Italian political scientist Piero Ignazi. He asked a simple but important question – why, in an era of mounting post-materialism and economic growth, was Europe witnessing a growing number of voters shifting over to parties like the Front National in France, the Austrian Freedom Party, and the Progress Parties in Scandinavia? He argued that alongside post-materialism a different mood had started to take root in Europe – this mood included the emergence of new priorities and issues not treated by the main parties, a disillusionment towards parties in general, a growing lack of confidence in the political system and its institutions, and a general pessimism about the future. It was anchored mainly in a reassertion of a different set of values – concerns about authority, patriotism, the role of the family and traditional moral values, much of which had been legitimised by the rise of neo-conservatism in the United States and Britain. The West, concluded Ignazi, was beginning to witness the emergence of a 'silent counter-revolution'.[26]

The values divide had probably always existed, but it was brought to the fore and increasingly politicised by the experiences of the 1970s and 1980s, in particular, the arrival of immigration as a major issue in

European politics and the perceived inability of established parties to respond in a way that satisfied increasingly anxious electorates. In the words of Ignazi, 'a mounting sense of doom, in contrast to post material-ist optimism, has been transformed into new demands, mainly unforeseen by the established parties. These demands include law and order enforce-ment and, above all, immigration control, which seems to be the leading issue for all new right-wing parties. This value change, stimulated by the reaction to postmaterialism and by a new combination of authoritarian issues, might be identified as a silent counter-revolution'.[27]

Those who flocked to national populists were significantly more likely to view immigrants as a burden on the welfare state, bad for the economy, and undermining national culture. Consider the Brexit vote. Among those who oppose gender equality, equality for same-sex couples, who want to see stiffer sentences for criminals and favour the death penalty, support for Brexit was higher than it was among liberals who hold the opposite views on these issues. The average Leave vote among people who want to see the death penalty reintroduced was 71 per cent, but among those who oppose the death penalty it was only 20 per cent.[28] Similarly, in Europe many studies of populist right-wing voters have now shown that the most important predictor of this support is their negative attitudes towards immigrants and ethnic minorities and their deep con-cern about how these are seen to threaten the nation state.

In a recent paper, Inglehart returned with his colleague Pippa Norris to explore the rise of the populist right and similarly came to the conclusion that it is a cultural backlash rather than economic insecurity that can account for the surge in votes for parties that present themselves as the main opponents to the rapid cultural changes that commenced in the 1960s. Even after applying social and demographic controls, all the cul-tural value scales were the strongest predictors of support. In their words: 'populist support was strengthened by anti-immigrant attitudes, mistrust of global and national governance and support for authoritarian values . . . Their greatest support is concentrated among the older generation, men, the religious, majority populations, and the less educated – sectors gener-ally left behind by progressive tides of cultural value change. The electoral success of these parties at the ballot box can be attributed mainly to their ideological and issue appeals to traditional values'.[29]

Nor is this divide simply situating young millennials against old white men. Rather, it is structured strongly by a sharp educational divide. In 2017, Marine Le Pen polled strongly among voters under 40 years old but who often had no university degree, while in countries like Austria national populists have polled strongest among younger men who have also lacked university education, the acquisition of which has been shown to encourage a more liberal outlook. Trump, too, made his strongest advances in areas of the United States where average education levels are low. Given that these deeper currents that run through educational experience and values are integral to the national populist revolt, it seems unlikely that the fourth wave will disappear anytime soon.

Conclusions

For these reasons, the new cultural divide between populist nationalists and cosmopolitan liberals will continue to push questions about national identity, immigration, and integration to the forefront. This development has been troubling for the mainstream but especially so for social democrats who have struggled to respond to the new debate in a language that resonates among their traditional working-class voters. Many of these voters are economically protectionist, hostile towards things like free trade and big banks, but they are also socially conservative, favouring restrictive immigration policies, adhering to more exclusive conceptions of nationhood, and instinctively sceptical of the shift towards transnational organisations (like the European Union). Crucially, and what many on the centre-left failed to recognise, is that their traditional voters often prioritise cultural protectionism over economic protectionism. Since 2015, in particular, the centre-left across Europe has suffered a loss of support. In 2017, social democrats in Austria, France, the Netherlands, and Germany experienced some of their worst performances on record, while national populists enjoyed some of their strongest.

A closely related trend concerns the steep decline in class voting that has also been taking place across most Western democracies in the post-war period. Though not restricted to the centre-left, the working classes are much less likely to be swayed by the old class-based loyalties that used to guide their parents and grandparents. By the 1990s, for example, it was estimated that social class voting was less than half as strong as it had been a generation earlier – thus issues like income and class became much weaker predictors of how somebody would vote at an election when compared with their attitudes towards issues like immigration and national identity. Many centre-left parties gambled that they could win over the new middle classes while retaining the traditional working classes but are now realising that they lost the gamble.

When seen in the whole, while national populism was a long time coming it is now difficult to avoid the conclusion that it will remain on the landscape for many years, perhaps decades, to come. While individual elections will come and go – Donald Trump may not be re-elected, Brexit may be overturned (although this seems unlikely) – the deeper currents that have underpinned these rebellions will remain firmly in place, swirling beneath our politics and no doubt waiting to find their expression in a new revolt against liberal democracy.

References and Further Reading

1 R. Eatwell and M.J. Goodwin, *National Populism: The Revolt against Liberal Democracy* (Penguin, 2018).

2 K. von Beyme (ed.), *Right-Wing Extremism in Western Europe* (Routledge, 1988).

3 V. Taylor, 'Social movement continuity: The women's movement in abeyance', *American Sociological Review*, vol. 54, issue 5 (Oct., 1989), pp. 761–75.

4 M.J. Goodwin, *New British Fascism: Rise of the British National Party* (Routledge, 2011). While von Beyme originally included the National Front in his 'third wave', I suggest the National Front, formed in 1967 and active mainly during the 1970s, is better situated in the second wave.

5 J. Rydgren, 'Is extreme right-wing populism contagious? Explaining the emergence of a new party family', *European Journal of Political Research*, vol. 44, issue 3 (May, 2005), pp. 413–37.

6 For further reading on this issue from The Darwin Lectures, see F. Halliday, 'Religious fundamentalism in contemporary politics', in P. Fara, P. Gathercole, and R. Laskey (eds.), *The Changing World* (Cambridge University Press, 1996), pp. 53–77 (the editors).

7 J.P. Zúquete, 'The European extreme-right and Islam: New directions?', *Journal of Political Ideologies*, vol. 13, issue 3 (Oct., 2008), pp. 321–44.

8 N. Spierings, M. Lubbers, and A. Zaslove, '"Sexually modern nativist voters": Do they exist and do they vote for the populist radical right?', *Gender and Education*, vol. 29, issue 2 (Jan., 2007), pp. 216–37.

9 H. Werts, P. Scheepers, and M. Lubbers, 'Euro-scepticism and radical right-wing voting in Europe, 2002–2008: Social cleavages, socio-political attitudes and contextual characteristics determining voting for the radical right', *European Union Politics*, vol. 14, issue 2 (Dec., 2012), pp. 183–205.

10 E. Carter, *The Extreme Right in Western Europe: Success or Failure?* (Manchester University Press, 2005); D. Art, *Inside the Radical Right: The Development of Anti-immigrant Parties in Western Europe* (Cambridge University Press, 2011).

11 Although that member would later defect to the breakaway party from the AfD, the 'Blue Party', which had been set up by his wife and former AfD leader Frauke Petry.

12 For example, P. Hainsworth (ed.), *The Extreme Right in Europe and the USA* (Pinter, 1992).

13 M.J. Goodwin and C. Milazzo, *UKIP: Inside the Campaign to Redraw the Map of British Politics* (Oxford University Press, 2015).

14 J. Mellon and G. Evans, 'Are Leave voters mainly UKIP?', British Election Study blog, www.britishelectionstudy.com/bes-impact/are-leave-voters-mainly-ukip-by-jonathan-mellon-and-geoffrey-evans/#.Wo6gahPFKWg (8 July 2016).

15 A. Amengay, A. Durovic, and N. Mayer, 'L'impact du genre sur le vote Marine Le Pen', *Revue française de science politique*, vol. 67, issue 6 (Dec., 2017), pp. 1067–87.

16 R. Ford and M.J. Goodwin, *Revolt on the Right: Explaining Support for the Radical Right in Britain* (Routledge, 2014).

17 G. King, O. Rosen, M. Tanner, and A.F. Wagner, 'Ordinary economic voting behavior in the extraordinary election of Adolf Hitler', *Journal of Economic History*, vol. 68, issue 4 (Dec., 2008), pp. 951–96.

18 'Europe's populist backlash', *New York Times*, 16 October 2013.

19 J. Palmer, 'The rise of the far right: A European problem requiring European solutions', *Guardian*, 15 November 2013.

20 *Supra* note 16.

21 M. Jacobs and M. Mazzucato, 'The Brexit-Trump syndrome: It's the economics stupid', LSE British Politics and Policy blog, http://blogs.lse.ac.uk/politicsand policy/the-brexit-trump-syndrome/ (accessed 24 July 2018).

22 D. Stockemer, 'The Economic Crisis (2009–2013) and electoral support for the Radical Right in Western Europe – some new and unexpected findings', *Social Science Quarterly*, vol. 98, issue 5 (Nov., 2017), pp. 1536–53.

23 E. Hernández and H. Kriesi, 'The electoral consequences of the financial and economic crisis in Europe', *European Journal of Political Research*, vol. 55, issue 2 (May, 2016), pp. 203–24.

24 D. Oesch, 'Explaining workers' support for right-wing populist parties in Western Europe: Evidence from Austria, Belgium, France, Norway, and Switzerland', *International Political Science Review*, vol. 29, issue 3 (June, 2008), pp. 349–73.

25 R. Inglehart, 'The silent revolution in Europe: Intergenerational change in post-industrial societies', *American Political Science Review*, vol. 65, issue 4 (Dec., 1971), pp. 991–1017.

26 P. Ignazi, 'The silent counter-revolution', *European Journal of Political Research*, vol. 22, issue 1 (July, 1992), pp. 3–34.

27 Ibid.

28 M.J. Goodwin and O. Heath, *Brexit Vote Explained: Poverty, Low Skills and Lack of Opportunities* (Joseph Rowntree Foundation, 2016).

29 R. Inglehart and P. Norris, 'Trump, Brexit, and the rise of populism: Economic have-nots and cultural backlash'. Paper for the roundtable on 'Rage against the Machine: Populist Politics in the U.S., Europe and Latin America', September 2016, annual meeting of the American Political Science Association, Philadelphia.

7 Extreme Longevity

SARAH HARPER

There have always been long-lived individuals. The challenge facing the twenty-first century is the sheer number of us and our children who are projected to survive to a century and more. As death has been pushed back across the life course, so that most people in high-income countries can expect to reach age 80 and over, so our societies and communities need to rethink our lives and the institutions that frame them. How did we attain such long lives? Will they be healthy or frail? Is there a maximum age a human can live to? And importantly, how can we ensure that current and future societies are able to maintain well-being across these long-lived lives, as well as equity within and between the generations?

Geert Adriaans Boomgaard was the world's first verifiable supercentenarian – someone who lives to 110 years. Yet there is little knowledge about his life. However, it is known that he was born and brought up in the low countries of Europe, that he was a foot soldier with Napoleon, and that he was born in 1788. He died in 1899, just before the turn of the twentieth century, and he held the record male life at 110 years and 135 days until 1966. Born just after Geert, in 1792 and dying in 1903 at 110 years and 321 days, making it into all three centuries (eighteenth, nineteenth, and twentieth), was Guernsey's Margaret Neve, the first recorded female supercentenarian and the second validated human to reach the age of 110.

There have always been long-lived individuals. There were probably only about ten centenarians in eighteenth-century Europe, though lack of verifiable data makes this difficult to estimate. Currently, the UK alone has about 15,000 centenarians, with significant growth seen in this age group over the past 30 years. Indeed, numbers doubled from 3,642 centenarians in 1986 to 7,750 in 2002, and then nearly doubled again to 14,910 by

2016 – a fourfold increase in 30 years. Notably, there are currently five female centenarians for each male. This growth in the numbers reaching old-old age is one of the drivers of the ageing of populations and a contributor to the increasing life expectancy of our populations. It is important to clarify the difference between *longevity* and *life expectancy*. While the first term refers to the longest individual human life, or maximum life span, life expectancy is the average number of additional years a person would live if current mortality conditions continued.[1]

As Geert demonstrated, however, we have always had long-lived people – those who have been able to push the boundaries back and reach extreme longevity. Madame Jeanne Calment from France, who died in 1997 aged 122 years, 164 days, currently has the longest verifiable human life span. The longest documented male life is that of Jiroemon Kimura of Japan, who died in 2013 aged 116 years and 54 days. These two individuals symbolise two current dynamics of life expectancy: women live longer than men, and French and Japanese populations are currently the longest lived.

There are four major questions framing the debate around extreme longevity:

1. Will increases in life expectancy continue? Will there be an increase in average years lived by humans?
2. Will life expectancy increase for all socio-economic groups? Will we all enjoy the benefits of longer lives or will only a select few?
3. Will advances in life expectancy be accompanied by advances in healthy life expectancy?
4. Will increases in life expectancy be accompanied by increases in life extension, or are we seeing a compression of longevity after 100? In other words, will the predicted increases in centenarians over the coming century be accompanied by increases in supercentenarians?

Will Increases in Life Expectancy Continue?

According to national statistics, in England, life expectancies slowly rose at the end of the nineteenth century, increasing significantly during the first half of the twentieth century, and then more steadily in the latter half of that century (Figure 7.1).[2] Furthermore, female life expectancy at birth

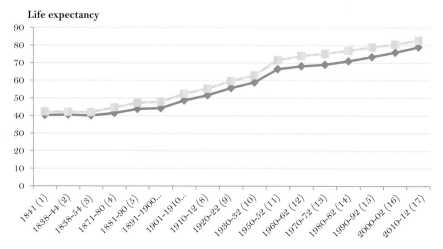

FIGURE 7.1 Period expectation of life at birth, 1841–2012
(data from Office for National Statistics)

in the longest-lived country at any time has increased each year since 1840 at a rate of approximately 2.5 years per decade, with a similar but slower increase for males.[3]

This steady change in life expectancy is associated with the epidemiological transition, characterised by a reduction in infectious and acute diseases and an increase in chronic and degenerative diseases. This is seen to comprise four distinct but overlapping stages: an age of pestilence and famine, with most people dying before the age of 40; an age of epidemics of infectious diseases, with most people surviving to their 50s; an age of chronic disease, with life expectancy reaching over 70; and an age of delayed degenerative diseases, with an increase in the age at which individuals succumb to such conditions, a lengthening of the time spent with chronic disease, and a fall in late-life death rates. Throughout the late nineteenth and twentieth centuries, many European and other high-income countries saw a steady reduction in mortality across the life course resulting in the rectangularisation of the life curve, whereby deaths have been consistently pushed further and further back, so that most deaths here now occur after age 80. This can be seen in the data for England (Figure 7.2).[4]

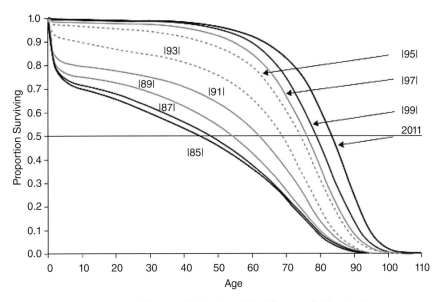

FIGURE 7.2 Rectangularisation of the life curve in England
(from S. Harper, *How Population Change Will Transform Our World* (Oxford University Press, 2016))

As Figure 7.2 shows, during the middle of the nineteenth century death still occurred across the life course. Steadily, however, death was pushed back into later and later ages, as infant and child mortality, maternal deaths, and adult deaths from infections and infectious diseases started to be reduced. Public health initiatives played a large role in these mortality declines.

From the middle of the nineteenth century onwards, mortality rates in many parts of Europe began to steadily improve. Between 1750 and 1850, England, France, and Sweden had gained less than 1 month of life expectancy per year. However, from 1850 to 1900, this more than doubled to 2 months per year driven by improvements in living standards and public health initiatives around sanitation, which enabled clean water, improved hygiene, and safe sewage disposal. Food availability and production also steadily improved across the century. Throughout the eighteenth and nineteenth centuries, trade, distribution, and transport systems transformed agriculture from a subsistence to a commercial enterprise, significantly increasing the availability of food. From the eighteenth century, new farming techniques enabled fresh meat to become available all year round, and improved transport allowed fresh fish to reach a greater proportion of the European population. At the same time, a wider variety of fruits and vegetables were grown, and bottling and canning techniques were successfully developed to preserve these throughout the year. These commercial initiatives were supported by government-backed public health moves that promoted the safe storage and handling of food. The overall result was that throughout the nineteenth century individual nutrition was considerably improved, resulting in a healthier European population better able to withstand the ravages of disease.

The large reduction in mortality across the twentieth century was primarily due to the conquering of infections and infectious diseases. The first three decades of the twentieth century saw a significant reduction in deaths with the introduction of effective pharmaceuticals, vaccinations, and antimicrobial chemotherapy, and the ability to identify new microbes. There were significant advances in diagnostic technology from the discovery of X-rays at the end of the nineteenth century through to ECGs and CAT and MRI scans. Geriatric medicine emerged in the mid-twentieth century with its recognition of the importance of co-morbidities of old age.[5]

A consideration of life expectancies in selected Organisation for Economic Co-operation and Development (OECD) countries this century reveals some clear patterns. Women have a clear advantage over men at all ages in terms of lower mortality risk. The longest-lived populations in terms of life expectancy at birth and from older ages (65 and 80) are Japanese and French women, currently with life expectancy at birth of 87 and 85.5 years, respectively, with a gap of some 3 years with UK women and 4 years with US women in life expectancy from birth (2015). This gap continues, and by age 80 French and Japanese women can expect to live on average another 11.6 years, while UK and US women can expect 9.6 years. A similar gap pertains to men and women in most countries. However, UK men have seen notable increases in life expectancy from birth and currently equal French men in this respect – with UK male life expectancies just under age 80 (79.2) (2015), possibly due to the significant decline in male smoking in the UK from the mid-1980s.

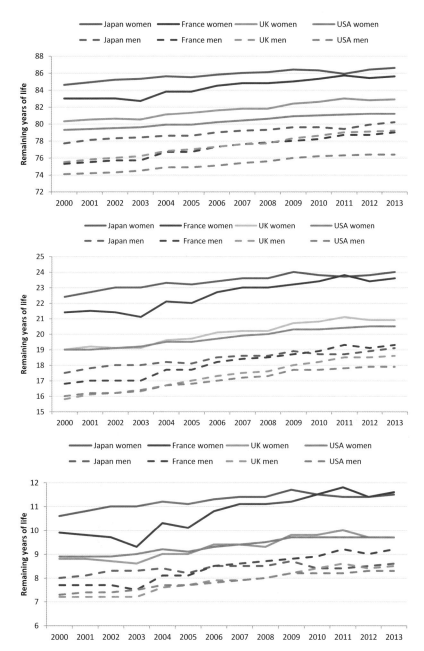

FIGURE 7.3 Life expectancy in selected countries at birth, age 65, and age 80
(data from the Organisation for Economic Co-operation and Development (OECD))

Looking to the future, models of life expectancy employing different modelling and different data all reveal continued increases. Data from the UK Office for National Statistics (ONS) for England and Wales, for example, predict that life expectancies at birth will reach 95.6 years for women and 93 years for men by the end of the century, and at age 65 years, life expectancies are expected to be 31.1 years for women and 29.9 years for males, giving ages of near 98 and 95, respectively.

Even higher ages are suggested by Christensen et al. based on modelling data from the Human Mortality Data Base, which predicts that the oldest age at which at least 50 per cent of the 2007 birth cohort will survive to in Japan is 107, half the 2007 babies born in the United States and France will live to 104 and 103, respectively, and in the UK those born in 2007 will make it to 103 (Figure 7.4).[6]

More recently, a study of World Health Organization (WHO) data by Kontis et al. suggests that South Korean women will have 35 years of life left by the time they reach age 65 (Figure 7.5).[7] Life expectancy is projected to increase in all 35 countries studied. There is a 90 per cent

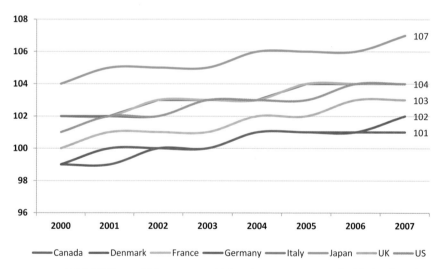

FIGURE 7.4 Oldest age at which at least 50 per cent of a birth cohort is still alive (cohorts 2000–2007)

(From K. Christensen, K.G. Doblhammer, R. Rau, and J.W. Vaupel, 'Ageing populations: The challenges ahead,' *The Lancet*, vol. 374, pp. 1196–202, Copyright 2009, with permission from Elsevier)

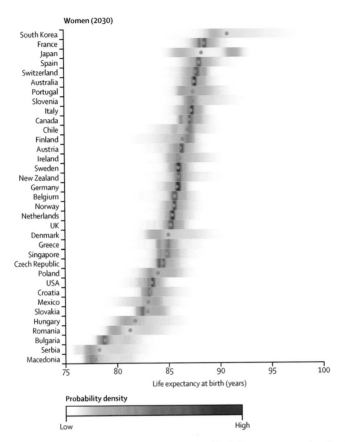

FIGURE 7.5 Female life expectancy at birth in 2030; posterior distribution of life expectancy and its median value (red dots show the posterior medians)[8] (From V. Kontis, J.E. Bennett, C.D. Mathers, et al., 'Future life expectancy in 35 industrialised countries: Projections with a Bayesian model ensemble', *The Lancet*, vol. 389 (2017), pp. 1323–35)

probability that life expectancy at birth among South Korean women in 2030 will be higher than 86.7 years, and a 57 per cent probability that it will be higher than 90 years. Projected female life expectancy in South Korea is followed by those in France, Spain, and Japan. There is a greater than 95 per cent probability that life expectancy at birth among men in South Korea, Australia, and Switzerland will surpass 80 years in 2030, and a greater than 27 per cent probability that it will surpass 85 years.

Such high life expectancies will lead to increasing numbers of people at extreme ages, including centenarians, as is reflected in the official

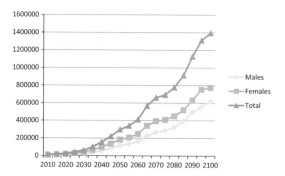

FIGURE 7.6 Projected numbers of centenarians in England and Wales to 2100[9]

(From G.W. Leeson, 'The impact of mortality development on the number of centenarians in England and Wales', *Journal of Population Research*, vol. 34, issue 1 (March, 2017), pp. 1–15. Data provided by Office for National Statistics)

forecasts of the ONS. In 2016, there were some 14,900 centenarians in the UK (12,480 females and 2,420 males).

As can be seen from Figure 7.6, this number is projected to increase steadily until the end of the twenty-first century, and the relative differ- ence between male and female numbers is expected to decrease. In 2010, there were five times as many female as male centenarians, but by 2100 numbers of male and female centenarians are expected to reach around 620,000 and 776,000, respectively: just 25 per cent more female than male centenarians. This corresponds to an absolute growth of almost 15,500 centenarians per year over the period from 2010 to 2100.

As Leeson also points out, the number of projected centenarians will again be determined by the complete demography of the future – i.e. the sizes of existing cohorts and developments in age-specific mortality as cohorts age, along with any migration.[10] Those projected to reach 100 years and more in 2100 will be the survivors of those aged at least 10 years in 2010. In particular, if age-specific later-life mortality declines more than projected in the official forecasts of the number of centenar- ians, these numbers will increase. From the 1960s to the late 1990s, the highest verified age at death in England and Wales ranged from 109 to 115 years. Projected death rates would mean that the highest ages being reached in the 2080s will be 116 to 123 years.[11] Leeson also considers a scenario in which projected mortality for those aged 55 years and over

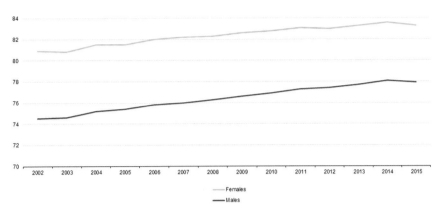

FIGURE 7.7 Life expectancy at birth in EU-28 slightly declined in 2015[12] (Source: Eurostat (online data code: demo_mlexpect))

decreases by some additional 5 per cent every 5 years in relation to the official projected mortality.[13] This scenario is based on the observed decreases in mortality for those aged 55 years and over in the previous 90 years from 1910–1919 to 2000–2009, when mortality of females aged 55–79 years declined on average every 5 years by between 3 and 5.5 per cent. For males, declines amounted to between 2.5 and 4 per cent. The number of centenarians in England and Wales by 2100 would then reach around 1.8 million.

However, it is worth noting that, from 2015 onwards, in both the EU and United States, increases in national life expectancy rates stalled (Figure 7.7). A variety of potential explanations have been given, from the EU-wide 2015 influenza epidemic to institutional factors such as the crisis in health care and social care.[14] However, two recent papers point out that for both England[15] and the United States[16] increases in inequality among the population may be having an impact on life expectancy progress at the national level.

Will Life Expectancy Increase for All Socio-economic Groups?

This leads us to the inequality question of whether life expectancy will continue to increase for all socio-economic groups, and whether we will all

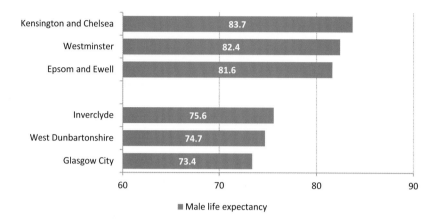

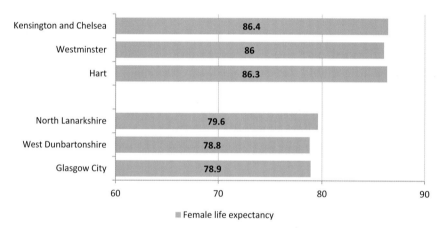

FIGURE 7.8 ONS regional life expectancies, 2014–2016
(Data from Office for National Statistics. Retrieved 25 February 2018 from www.statistics.gov.uk/
StatBase/Product.asp?vlnk=8841&Pos=1&ColRank=1&Rank=272)

enjoy the benefits of longer lives or they will be for only a few. In terms of
inequalities in life expectancy, it was originally thought that social class
differentials in mortality were associated mainly with poverty measures.
More recently, the focus has been on the social gradient in mortality risk,
whereby lower income groups within a society have a higher mortality
rate, despite being well above the poverty line. Socio-economic status
rather than poverty has now become the central concept for investigating
social inequalities in mortality. It is well known at the national level that

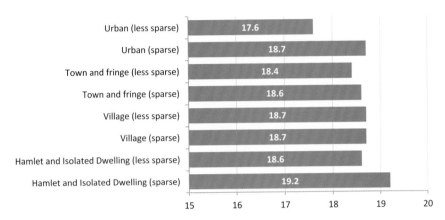

FIGURE 7.9 ONS male life expectancies from age 65 by urban–rural residence (Data from Office for National Statistics)

there are considerable inequalities in life expectancy between those individuals living in affluent and non-affluent areas (Figure 7.8), and also between urban and rural regions (Figure 7.9), as English ONS data for England and Wales illustrate.[17]

In addition, a study by Harper et al. revealed that even among a more affluent subset – those individuals fortunate to be in receipt of an occupational pension – there were considerable differentials for both men and women. Analysis of 500,000 individuals who had died when in receipt of occupational pensions revealed considerable variation in life expectancy based on variation in income, occupation, and health (Table 7.1).[18]

Retiring in 'normal health' can add up to 3.5 years of extra life than retiring in 'ill-health'. Of most importance is health across the life course – all else being equal, there is a difference of up to 4 to 5 years in life expectancy between the least healthy and healthiest lifestyles. Income has an impact on life expectancy to lifestyle, although the effect is different for men and women. Men with a history of high salaries have a life expectancy of 3 years longer than those with the lowest salaries, but the effect of personal income is smaller for women than for men. Occupation – whether an individual has carried out a 'manual' or 'non-manual' role – can account for up to a 0.75 year difference in life expectancy for men and up to 1.5 years for women. Manual workers tend to have shorter life expectancies.

Table 7.1 *Range of life expectancies from age 65 for different combinations of affluence, lifestyle, retirement health, and occupation*

	Lowest life expectancy	Highest life expectancy	Difference
Men	12.0 years	23.0 years	11.0 years
Women	13.9 years	23.5 years	9.6 years

Male

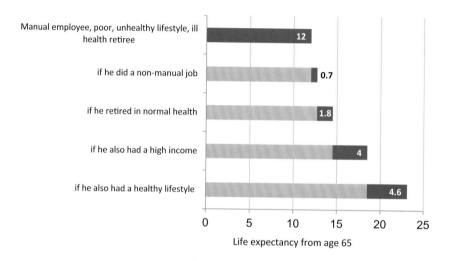

Female

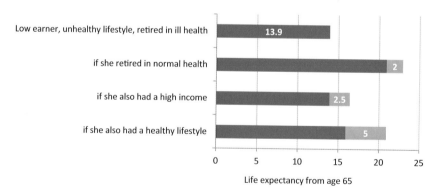

FIGURE 7.10 Impact of different factors on longevity[19]
(From S. Harper, K. Howse, and S. Baxter, *Living Longer and Prospering? Designing an Adequate, Sustainable and Equitable UK State Pension System* (Club Vita LLP and Oxford Institute of Ageing, 2011))

As Figure 7.10 reveals, taking the example for men, at the top we have our unhealthy group (manual employee, poor, unhealthy lifestyle, ill-health retiree): only 12 years at age 65. If they had done a non-manual job, they would have added 0.7 years; retired in normal health, 1.8 years; had a high income but unhealthy lifestyle, 4 years. It was having a healthy lifestyle across one's life that would add on 4.6 years, so health has a huge impact on life expectancy and it needs to be across our lives.

Importantly, as Figure 7.11 shows, it is not just life expectancy but healthy life expectancy that also varies considerably between social groups. The figure illustrates the life expectancy and healthy life expectancy of English men in 2014 by the area in which they live, across ten areas. There is a striking contrast between those living in the most and the least deprived areas. Those in the most deprived areas can at age 65 expect to reach 80, but with only just over a third of this time, 6 years, in good health. Those living in the least deprived areas expect on average

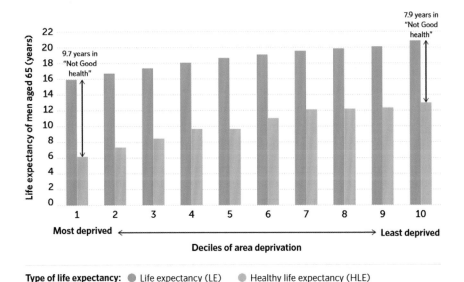

FIGURE 7.11 Life expectancy and healthy life expectancy
(Source: *Future of an Ageing Population*, 2016, Government Office for Science report, www.ageing.ox.ac.uk/publications/view/328)

to live until they are 86 and will enjoy over half these 21 years (13 years) in good health.[20]

Will Advances in Life Expectancy Be Accompanied by Advances in Healthy Life Expectancy?

It may be argued that from society's perspective the key question concerning increasing life expectancies is whether these will be healthy or disabled and frail extra years. Additional healthy years enable individuals to continue to contribute to their communities – whether through economic activity, caring, or volunteering. If these healthy years increase as a proportion of overall life expectancy, there may also be a reduction in years spent in frailty, reducing both individual burden as well as that shared by families, friends, and society in late life care. If the additional years are in ill health, then the social and public health burdens will increase considerably.

In terms of the relationship of life expectancy to healthy life expectancy, there are three main theoretical stances: *compression of morbidity*, in which healthy life expectancy increases faster than life expectancy, and extra years of life are healthy ones; *expansion of morbidity*, in which healthy life expectancy does not keep pace with life expectancy, and the extra years of life are unhealthy ones; and *dynamic equilibrium*, in which disabled life expectancy is increasing, but the severity of ill health is reducing. A general conclusion is that we are pushing back the onset of frailty but maintaining that state for longer. In particular, the modern drivers of longevity, science and technology, are enabling those with frailty to live longer in that state.

There is thus growing evidence that increases in healthy life expectancies are not keeping pace with gains in life expectancy in many high-income countries, particularly at older ages. As with life expectancy, variations in health and disability are closely related to socio-economic differences. Surveys of the European population have increased understanding of health inequalities, suggesting that inequalities in healthy life years at age 50 across Europe are increasing, partly explained by levels of material deprivation and long-term unemployment.[21]

It should be noted that there is a major limitation in the lack of comparable measures of healthy life expectancy measures. Recently, however, the Global Burden of Disease programme, which estimates healthy life expectancy for 187 countries, is providing comparative data;[22] the introduction in the European Union of healthy life years, a Disability-Free Life Expectancy measure, has allowed comparative work among EU countries; and the Global Activity Limitation Indicator (GALI) is attempting to measure restrictions in participating in social and economic roles, as envisaged in the WHO International Classification of Functioning, Disability and Health.[23]

Adding a further dimension to the debate is the proposition that increasing chronic disease across the life course in many high-income countries will even reduce life expectancy. As Olshansky et al. posited in the *New England Journal of Medicine*, 'We anticipate that as a result of the substantial rise in the prevalence of obesity and its life-shortening complications such as diabetes, life expectancy at birth and at older ages could level off or even decline within the first half of this century.'[24] There is, however, now growing evidence that obesity will increase years lived with disability, rather than reduce life expectancy per se. Indeed, while smoking in particular reduces life expectancy, obesity leads to more disabled years. Obesity is rarely fatal but leads to a range of chronic conditions, whereas smoking is directly associated with usually fatal conditions such as cardiovascular disease, stroke, and lung cancer. Research by Klijs et al. on 6,446 individuals from the Dutch Permanent Survey of the Living Situation (POLS) to compare the effect of body mass index (BMI), smoking, and alcohol consumption on life expectancy with disability suggested that, compared with smoking and drinking alcohol, obesity is most strongly associated with an increased risk of spending many years of life with disability. While obesity reduced life expectancy by only 1.4 years, compared with smoking (4 years) and alcohol (3 years), it increased the years living with disability by 5.9 years, significantly more than smokers (3.8 years) and drinkers (3.1 years).[25] Similarly, a study of over 8,000 older adults (over age 50 years) from the Mexican Health and Aging Study (MHAS) found that obesity had high association with increased disability but a low association with increased morbidity.[26] Studies in the United States have found similar results.[27]

Will Increases in Life Expectancy Be Accompanied by Increases in Life Extension?

While there is a body of evidence indicating that lives will continue to be extended,[28] this is also contested by those who argue that there is a maximum to human life spans.[29]

Key to this question is whether the predicted increases in centenarians over the coming century will be accompanied by increases in super-centenarians, or whether what we are seeing is a compression of longevity after 100.[30] Contrary to the expectations of some analysts in the 1980s, mortality reductions in the oldest-old have shown no signs of slowing down in recent years. Two key questions to ask of these data are whether they show any signs of a compression of mortality and whether mortality rates at older ages continue to rise exponentially. One way to consider this is to search for evidence of a compression of mortality in countries with the highest life expectancy, as this would strongly suggest that we are approaching an upper limit to human longevity in the developed world. When the modal age of death is itself increasing, a continuous decrease in the standard deviation of the age distribution of deaths above the mode would indicate a compression of mortality. If there is no change in this particular measure of dispersion, so that the whole distribution of deaths above the mode is sliding proportionally to older ages, then we have evidence for the so-called shifting mortality scenario.[31]

We now have sufficient evidence from both Japan and France, our two longest-lived populations, to suggest that life expectancy at very old ages is continuing to increase. Work by Cheung and Robine analysing Japanese mortality from 1950 to 2000 (Figure 7.12) showed that not only had there been a strong and linear increase in the modal age at death over the previous 50 years, but also the standard deviation of ages at death above the mode had stopped decreasing in the mid-1980s for women and in the 1990s for men. In other words, data from the country with the lowest mortality in the world showed no sign of a compression of mortality.[32]

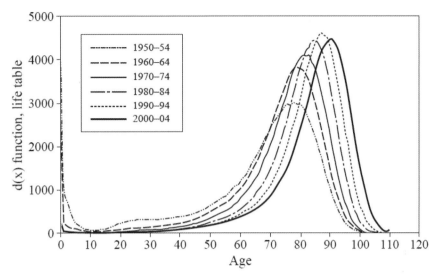

FIGURE 7.12 Changes in the distribution of ages at death for women in Japan from 1950–1954 to 2000–2004
(From S.L.K. Cheung and J.-M. Robine, 'Increase in common longevity and the compression of mortality: The case of Japan', *Population Studies*, vol. 61, issue 1 (March, 2007), pp. 85–97)

The Future?

On 15 April 2017, the last living link to the nineteenth century came to an end. Emma Morano, who was born in November 1899 and is believed to have been the last survivor from the nineteenth century, died at her home in northern Italy aged 117. One of the five longest-lived people in recorded history, she was born in Civiasco in the Piedmont region of northern Italy and lived through two world wars and more than 90 Italian governments. Emma is succeeded by Violet Brown, a 117-year-old Jamaican woman. The second- and third-oldest people are two Japanese women, Nabi Tajima and Chiyo Miyako, who were born, respectively, in August 1900 and May 1901. Ana Vela, 115, a Spanish woman, was born in October 1901 and is currently the oldest European and the fourth-oldest person in the world.

Our understanding of the genetic drivers of very long-lived people is being enhanced by the increasing numbers of these long-lived individuals.

Robine and colleagues' study of extremely old people in Japan observed that, despite being on average over 10 years older, the long-lived participants were healthier than the controls. They possessed significantly better biological and physiological risk factor profiles, less age-related disease, and better physical and cognitive function. The study thus endorsed the concept of a 'healthy ageing' phenotype, whereby certain individuals are somehow able to delay or avoid major clinical disease and disability until late in life.[33] Other studies have revealed that supercentenarians, those surviving 110 years or more, display an exceptionally healthy ageing phenotype, supporting the delay of major chronic diseases and disabilities. Such individuals had little cardiovascular disease and reported no history of cancer or diabetes.[34]

To date, the reduction in mortality and morbidity at the population level has been primarily driven by healthy living and disease prevention and cure. Indeed, we are still considering how much life expectancy we may gain without the intensive application of scientific medicine. The main approach has been to estimate the impact on life expectancy of delaying the onset of what we know to be age-related diseases rather than eliminating them altogether. However, some 30 years ago Olshansky et al. attempted to project the potential effect of eliminating certain major causes of death altogether. They estimated that hypothetical cures for all circulatory diseases, diabetes, and cancer would increase life expectancy at birth in the United States by 15.8 years for females and 15.3 years for males (beyond 1985 levels) – equivalent to a 75 per cent reduction in mortality from all causes.[35]

However, we are now at the cusp of estimating how much life expectancy we might expect to gain with the intensive application of scientific medicine – with the acknowledgement that advances across the twenty-first century of regenerative medicine and age-retardation are likely. From promoting longevity through dietary restriction to pluripotent stem cell research, the potential of these therapies to increase individual life spans seems significant.[36]

Take stem cell therapy, for example, and the ability of pluripotent cells to develop into skin, nerve, muscle, or practically any other cell type. Given that increases in life expectancy have been accompanied by the rising incidence of chronic and degenerative diseases, induced pluripotent stem (iPS) cells may address this challenge. Such cells have now become an important tool, whereby with gene-editing technologies they have become valuable in the fields of human development and neurological diseases. For example, iPS cells have been used to create organoids such as mini-guts and mini-livers, and research has generated cells with precise combinations of Alzheimer's-associated mutations in order to study the effects. The European Bank for Induced Pluripotent Stem Cells, currently centred in Cambridge, UK, has launched a catalogue of stand-ardised iPS cells for use in disease modelling. Key advances are also being made in drug discovery, as particular iPS cells provide a plentiful source of patient-derived cells to screen or test experimental drugs. And in the laboratory, there have been successes at growing mouse organs in rats, though it seems we are still a long way from growing human organs in larger animals such as pigs as currently the resulting human-animal chimeras don't grow well, and few human cells survive.[37] However, for many in this scientific community, it is 'the availability of therapeutic cell types for the replacement of diseased or worn-out tissues and organs that remains the ultimate goal of regenerative medicine and provides a tantal-ising glimpse into a future of personalised medicine tailored to an ageing population'.[38]

In 2013, iPS cells were developed from the skin cells of people with age-related macular degeneration, which in one case at least halted the macular degeneration and brightened vision. However, the question of whether such therapy also induces tumour growth halted these trials. In addition, it is now recognised that most reprogramming techniques are inefficient: only a small fraction of cells end up fully reprogrammed. Each new advance brings with it new challenges – the risk of increased cancer with stem cell therapy, for example.

What Kind of Society Will Emerge with Extreme Longevity?

It has recently been posited that we should recognise the degenerative ageing process itself as a medical problem to be addressed.[39] Such recognition, it is argued, may accelerate research and development: the scientific community will further engage with ageing; there may be a call by the general public to demand preventive therapies; the pharmaceutical and medical technology industries will increase their development and marketing of effective ageing-ameliorating therapies and technologies; health insurance, life insurance, and health care systems will obtain a new area for reimbursement practices, which would lead to the promotion of healthy longevity; regulators and policy makers will be encouraged to prioritise and increase public investment into ageing-related research and development.

There is, however, an opposing view, which argues that ageing and the inevitability of eventual death are part of the human condition; finality is an essential component of that which makes us human. As Cicero wrote over 2,000 years ago:

> Men are a product of nature and the virtuous life follows the ways of the plants, we grow and flourish, then ripen, wither and fall. There is a time for each.[40]

Even if we do not achieve the acclaimed goal of near infinity in life spans, nor even come close with life expectancies of whole populations over age 120 – currently, remember, the longest-lived individual – extending our life spans will have significant economic, social, and political consequences. Recent projections by Leeson, for example, suggest that England and Wales will have more than 1,800 individuals aged over 110 by the middle of the century, and over 80,000 by the end. Indeed, if current trends in life expectancy continue, by the close of the twenty-first century some 350 individuals in England and Wales alone will achieve the longevity of Madame Calment at over 120.[41]

Current major concerns include public spending on pensions, high dependency ratios between workers and non-workers, increases in health care costs, declining availability of family-based care, and a slowdown in consumption due to an increase in older people and a decrease in younger people. Some of these fears are supported by evidence, but many are speculative myths, widespread in public debate but lacking a robust evidential base. They are generally based on assumptions developed from the characteristics and behaviours of current older populations. It is likely, for example, that future generations of older adults will have higher levels of human capital – in terms of education, skills, and abilities – and that old age, as defined by retirement and dependency, will occur at far older ages than currently. In addition, these are all issues that can be addressed by policy, given the political and economic will.

Alternatively, it may be argued that if healthy life expectancy keeps up with life expectancy, older adults will simply remain active and contributory for longer – well into their 80s and 90s perhaps. While that would indeed solve many of the concerns raised above, it would then raise a second broad set of issues. As is discussed elsewhere,[42] the key challenge for both current and future ageing societies is how to maintain well-being across the life course, and within and between generations, as age structures change leading to significant implications for the redistribution of national and global resources. For in trying to maintain population well-being, societies that have a growing proportion of their populations in old age must consider how to redistribute resources within this new demographic. In other words, there will be a need to move resources away from a focus on younger people towards older people in an equitable manner, both inter-generationally (between the generations) and intra-generationally (within each generation).

At risk, for example, is the *generational contract*. In most societies there exists a strong exchange of social and economic support between the generations. Parents look after their dependent children, and, when they grow up, these adult children care for their dependent parents. In traditional societies, this is done within families and households, in modern societies via public welfare systems. Given their ageing populations, many countries are now debating whether this contract should now be adapted so that older adults bear more direct responsibility for their own late life care and support.

A similar question concerns *generational succession*. Within most societies there is a clear concept of the generational transmission of status, assets, power, and wealth down through the generations at a steady rate. What will happen when, due to the extreme longevity of a population, individuals do not inherit from their parents until they, the children, are in their 80s? Or from their grandparents until they, the grandchildren, are in their 80s? What happens to our workforces when healthy, active individuals are still in full employment in their ninth decade? To our communities? Our societies? Our polities?

In addition, as described elsewhere,[43] we are already seeing the *ageing of life transitions*: increase of age at first marriage and at remarriage, at leaving the parental home, at first childbirth. While public and legal institutions may be lowering the age threshold into full legal adulthood, individuals themselves are choosing to delay many of those transitions that demonstrate a commitment to full adulthood – full economic dependence from parents, formal adult union through marriage or committed long-term cohabitation, and parenting.

Such change is, of course, possible to accommodate, but not without both an adaptation of our institutions and a rethinking of the life course. This requires a move away from one in which there is a clear progression along a linear line in terms of work and income to a more fluid life course – a life course that is not segmented into education, work, retirement, but follows a more flexible overlapping path. One, for example, in which education and retraining continue throughout our lives, and parents are able to withdraw from full-time employment to care for young children, returning in their late 40s, refreshed and ready for 30 more years of economic employment.

Approaching a Century and Beyond

Let us end with reflections on a long life by Oliver Sacks:

> *I begin to feel, not a shrinking but an enlargement of mental life and perspective. One has had a long experience of life, not only one's own life, but others', too. One has seen triumphs and tragedies, booms and busts, revolutions and wars, great achievements and deep ambiguities, too ... One is more conscious of transience and, perhaps, of beauty ... I can imagine, feel in my bones, what a century is like, ... I think of old as a time of leisure and freedom, freed from the factitious urgencies of earlier days, free to explore whatever I wish, and to bind the thoughts and feelings of a lifetime together.* [44]

To feel in one's bones what a century of life is like – that is a gift which all who are born should enjoy, 100 years of healthy, happy, active longevity.

Notes, References, and Further Reading

1 Technically, the definition is the average number of additional years a person would live if the mortality conditions implied by a particular life table applied.
2 Office for National Statistics, 'Estimates of the very old (including centenarians): 2002 to 2016', *Statistical Bulletin* (2017).
3 J. Oeppen and J.W. Vaupel, 'Broken limits to life expectancy', *Science*, vol. 296, issue 5570 (May, 2002), pp. 1029–31; R.G. Westendorp, 'Are we becoming less

disposable? Evolution has programmed us for early survival and reproduction but has left us vulnerable to disease in old age. In our present affluent environment, we are better adapting to these improved conditions', *EMBO Reports*, issue 5 (Jan., 2004), pp. 2–6.

4 S. Harper, *How Population Change Will Transform Our World* (Oxford University Press, 2016).

5 For further reading on this topic from The Darwin Lectures, see L. Borysiewicz, 'Plagues and medicine', in J.L. Heeney and S. Friedemann (eds.), *Plagues* (Cambridge University Press, 2017), pp. 66–91, and C. Kenyon and C. Cockroft, 'Surviving longer', in E. Shuckburgh (ed.), *Survival* (Cambridge University Press, 2008), pp. 178–204 (the editors).

6 K. Christensen, G. Doblhammer, R. Rau, and J.W. Vaupel, 'Ageing populations: The challenges ahead', *The Lancet*, vol. 374, issue 9696 (Oct., 2009), pp. 1196–208.

7 V. Kontis, J.E. Bennett, C.D. Mathers, et al., 'Future life expectancy in 35 industrialised countries: Projections with a Bayesian model ensemble', *The Lancet*, vol. 389, issue 10076 (Feb., 2017), pp. 1323–35.

8 Ibid., p. 1327.

9 G.W. Leeson, 'The impact of mortality development on the number of centenarians in England and Wales', *Journal of Population Research*, vol. 34, issue 1 (March, 2017), pp. 1–15.

10 G.W. Leeson, 'Future prospects for longevity', *Post Reproductive Health*, vol. 20, issue 1 (March, 2014), pp. 11–15.

11 G.W. Leeson, 'Increasing longevity and the new demography of death', *International Journal of Population Research*, vol. 2014, article ID 521523 (June, 2014), pp. 1–7; R. Thatcher, 'The demography of centenarians in England and Wales', *Population: An English Selection*, vol. 13, issue 1 (2001), pp. 139–56.

12 Eurostat Database, Cross-cutting-topics – Quality of life – Health – Health status – Life expectancy by age and sex (data code: demo_mlexpect), http://ec.europa.eu/eurostat/ (accessed 31 July 2018).

13 G.W. Leeson, 'The impact of mortality development on the number of centenarians in England and Wales', *Journal of Population Research*, vol. 34, issue 1 (March, 2017), pp. 1–15.

14 R. Loopstra, M. McKee, S.V. Katikireddi, et al., 'Austerity and old-age mortality in England: A longitudinal cross-local area analysis, 2007–2013', *Journal of the Royal Society of Medicine*, vol. 109, issue 3 (2016), pp. 109–16.

15 M. Fransham and D. Dorling, 'Have mortality improvements stalled in England?', *The BMJ*, vol. 357 (May, 2017).

16 S. Harper, J.S. Kaufman, and R.S. Cooper, 'Declining US life expectancy: A first look', *Epidemiology*, vol. 28, issue 6 (Nov., 2017), pp. e54–56.

17 Office for National Statistics, www.statistics.gov.uk/StatBase/Product.asp?vlnk=8841&Pos=1&ColRank=1&Rank=272 (accessed 25 February 2018; no longer accessible).

18 S. Harper, K. Howse, and S. Baxter, *Living Longer and Prospering? Designing an Adequate, Sustainable and Equitable UK State Pension System* (Club Vita LLP and Oxford Institute of Ageing, 2011).

19 Ibid. The effects described above are the independent and in the aggregate – a spread in life expectancies of over 11 years for men and almost 10 years in women.

20 Government Office for Science, *Future of an Ageing Population* (Crown Copyright, 2016).

21 C. Jagger, C. Weston, E. Cambois, et al., 'Inequalities in health expectancies at older ages in the European Union: Findings from the Survey of Health and Retirement in Europe (SHARE)', *Journal of Epidemiology and Community Health*, vol. 65 (April, 2011), pp. 1030–35; T. Fouweather, C. Gillies, P. Wohland, et al., 'Comparison of socio-economic indicators explaining inequalities in Healthy Life Years at age 50 in Europe: 2005 and 2010', *The European Journal of Public Health*, vol. 25, issue 6 (Dec., 2015), pp. 978–83.

22 J.A. Salomon, H. Wang, M.K. Freeman, et al., 'Healthy life expectancy for 187 countries, 1990–2010: A systematic analysis for the Global Burden of Disease Study 2010', *The Lancet*, vol. 380, issue 9859 (Dec., 2012), pp. 2144–62.

23 N. Berger, H. Van Oyen, E. Cambois, et al., 'Assessing the validity of the Global Activity Limitation Indicator in fourteen European countries', *BMC Medical Research Methodology*, vol. 15, issue 1 (Jan., 2015).

24 S.J. Olshansky, D.J. Passaro, R.C. Hershow, et al., 'A potential decline in life expectancy in the United States in the 21st century', *New England Journal of Medicine*, vol. 352, issue 11 (March, 2005), pp. 1138–45, 1142.

25 B. Klijs, W.J. Nusselder, C.W. Looman, and J.P. Mackenbach, 'Contribution of chronic disease to the burden of disability', *PLoS One*, vol. 6, issue 9 (Sep., 2011), p. e25325.

26 A. Kumar, A.M. Karmarkar, A. Tan, et al., 'The effect of obesity on incidence of disability and mortality in Mexicans aged 50 years and older', *Salud Publica de Mexico*, vol. 57, issue 1 (July, 2015), pp. s31–38.

27 S. Al Snih, K.J. Ottenbacher, K.S. Markides, et al., 'The effect of obesity on disability vs mortality in older Americans', *Archives of Internal Medicine*, vol. 167, issue 8 (April, 2007), pp. 774–80.

28 *Supra* notes 4, 6, 10 (second reference); J. Bongaarts, 'How long will we live?', *Population and Development Review*, vol. 32, issue 4 (Nov., 2006), pp. 605–28; J.-M. Robine, Y. Saito, and C. Jagger, 'The emergence of extremely old people: The case of Japan', *Experimental Gerontology*, vol. 38, issue 7 (July, 2003), pp. 735–39; J.W. Vaupel, 'Biodemography of human ageing', *Nature*, vol. 464, issue 7288 (March, 2010), pp. 536–42; J.R. Wilmoth and J.-M. Robine, 'The world trend in maximum life span', *Population and Development Review*, vol. 29 (2003), pp. 239–57.

29 X. Dong, B. Milholland, and J. Vijg, 'Evidence for a limit to human lifespan', *Nature*, vol. 538, issue 7624 (Oct., 2016), pp. 257–59; S.J. Olshansky, B.A. Carnes, and A. Désesquelles, 'Prospects for human longevity', *Science*, vol. 291, issue 5508 (Feb., 2001), pp. 1491–92.

30 S. Harper and K. Howse, 'An upper limit to human longevity?', *Journal of Population Ageing*, vol. 1, issue 2–4 (2008), pp. 99–106.

31 *Supra* note 28 (first reference).

32 S.L.K. Cheung and J.-M. Robine, 'Increase in common longevity and the compression of mortality: The case of Japan', *Population Studies*, vol. 61, issue 1 (March, 2007), pp. 85–97.

33 *Supra* note 28 (second reference).

34 D.C. Willcox, B.J. Willcox, N.C. Wang, et al., 'Life at the extreme limit: Phenotypic characteristics of supercentenarians in Okinawa', *The Journals of Gerontology. Series A, Biological Sciences and Medical Sciences*, vol. 63, issue 11 (Nov., 2008), pp. 1201–8; B.J. Morris, T.A. Donlon, Q. He, et al., 'Genetic analysis of TOR complex gene variation with human longevity: A nested case–control study of American men of Japanese ancestry', *The Journals of Gerontology. Series A, Biological Sciences and Medical Sciences*, vol. 70, issue 2 (Feb., 2015), pp. 133–42.

35 S.J. Olshansky, B.A. Carnes, and C. Cassel, 'In search of Methuselah: Estimating the upper limits to human longevity', *Science*, vol. 250, issue 4981 (Nov., 1990), pp. 634–40.

36 L. Fontana and L. Partridge, 'Promoting health and longevity through diet: From model organisms to humans', *Cell*, vol. 161, issue 1 (March, 2015), pp. 106–18.

37 G. Vogel, 'Human organs grown in pigs? Not so fast', *Science*, 26 January 2017, www.sciencemag.org/news/2017/01/human-organs-grown-pigs-not-so-fast (accessed 31 July 2018).

38 P. Fairchild, *Tapping the Fountain of Youth: Exploiting Stem Cells for the Treatment of Degenerative Disease* (Oxford Institute of Population Ageing Seminar, 2017).

39 I. Stambler, 'Recognizing degenerative aging as a treatable medical condition: Methodology and policy', *Aging and Disease*, vol. 8, issue 5 (Sep., 2017), pp. 583–89.

40 M.T. Cicero, *Cato Maior de Senectute* [On old age], 44 BC.

41 *Supra* note 11.

42 *Supra* note 4.

43 S. Harper, *Ageing Societies: Myths, Challenges and Opportunities* (Hodder Arnold, 2006).

44 O. Sacks, 'The joy of old age. (no kidding.)', *New York Times*, 6 July 2013, www.nytimes.com/2013/07/07/opinion/sunday/the-joy-of-old-age-no-kidding .html (accessed 31 July 2018).

8 Extremes of Power in the Universe

ANDREW C. FABIAN

The Universe contains many objects that are extreme in some way. Here we shall look at extremes of power, where power is defined as the rate at which energy is released. Extreme power means a very large amount of energy released on a very short timescale, either all at once or continuously.

We begin by noting that there *is* an upper limit to power and we will call that the ultimate power. It has a value of about c^5/G.[1] The speed of light, c, is very large and G (Newton's gravitational constant) is small, so c^5/G is an enormous number. It corresponds to 100 thousand billion trillion times (10^{26} x) the steady power of the Sun (10^{26} L_{sun}), or 10 thousand times the power of all the stars in all the galaxies in the observable Universe (10^{22} L_{sun})! Astronomically extreme! For comparison, a 4W cycle lamp has a luminous power about 10^{22} times fainter than that of the Sun, i.e. 10^{-22} L_{sun}.

Nevertheless, in the past few years powerful distant events have been observed within a factor of about 10 of the ultimate value. They involve mergers of neutron stars and black holes and are discussed in more detail later.

You may be concerned that the maximum power formula does not include the mass of the object. How does that work? Energy is conserved, since the length of time over which the power lasts is proportional to the mass consumed, so if the power comes from a black hole with a mass of a billion Suns, it lasts a billion times longer than if it is from a neutron star of solar mass.

We start looking at individual events by considering the Sun, which, fortunately for us, is seen as a mostly steady source of energy with a power L_{sun}. On closer inspection, however, sunspots and other activity

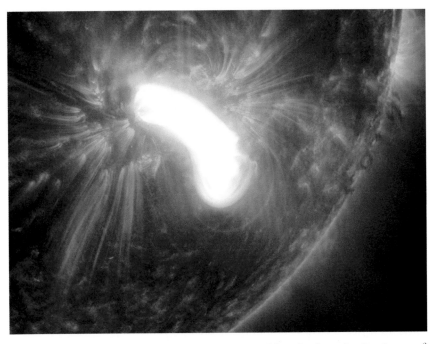

FIGURE 8.1 Extreme ultraviolet image of the solar flare of 26 October 2014[2]

can be seen. These are produced by magnetic fields wound up by the differential rotation of the Sun. One rotation of the Sun takes 26 days at the equator and about 30 days at the poles. Magnetic energy is stored and grows as the fields are tangled up. They then burst through the surface, creating twisted loops of field that release their energy by accelerating charged particles to X-ray-emitting energies: an event called a solar flare (Figure 8.1). At the same time, parts of the outer atmosphere of the Sun, the corona, can be expelled in an event called a coronal mass ejection, which leads to energetic clouds of energetic particles flooding out towards the planets.

The most energetic solar flare observed so far was witnessed independently by R. Carrington and R. Hodgson from London in 1859. The instruments at Kew Magnetic Observatory were simultaneously disturbed to a great extent. Early the following morning, a magnetic storm was recorded worldwide. The coronal matter ejected during what is now known as the 'Carrington event' headed straight for Earth, and collided with its magnetic field, triggering bright aurorae seen in the Caribbean and inducing currents in telegraph wires, which literally shocked their operators.

Nothing on this scale has headed directly for Earth since, although an event in 2012 witnessed by a solar-orbiting observatory was probably comparable, aimed away from the Earth. A much smaller solar-induced magnetic storm in March 1989 caused the province of Quebec's electrical grid to go offline for 12 hours. Should a Carrington-scale event recur and hit the Earth, it could disable many of the satellites on which we now rely for navigation and communications. The currents induced on Earth could ruin power transmission grids: a costly catastrophe that some estimates put at trillions of dollars.

We cannot yet predict when one might happen, but we can study X-ray superflares detected from stars similar to the Sun (Figure 8.2). Superflares are thousands of times more energetic than the Carrington event but show a power-law distribution of energies that extrapolates to match that of solar flares. Flares with the energy of the Carrington event occur every few decades and of those only 10 per cent pass near the Earth so we experience one every few hundred years. Superflares may occur every few million years for an object like the Sun. Let's hope that we never observe one! Even solar flares at the level of the Carrington event pose a major risk, e.g. for interplanetary travel by humans.

Although solar flares are potentially dangerous, they are not very powerful events compared with the wider Universe. Their rate of energy release, or power, is less than one thousandth of the steady power output of the Sun.[3]

Let's leave the Solar System and look for far more powerful objects and events. Persistently luminous objects held together by their own self-gravity, like the Sun and other stars, are limited by radiation pressure. Radiation produced in a star interacts with the gas of the star as it scatters its way out, exerting a pressure on the gas, pushing it outwards. Even beyond stars, radiation is scattered and absorbed by tiny grains of dust, which are (relatively) common in space. In the interstellar space between the stars in our galaxy, about 1 per cent of the mass is in such tiny grains, with the rest of the mass being in gas. Comets reveal the interaction between sunlight and dust by creating long tails, which

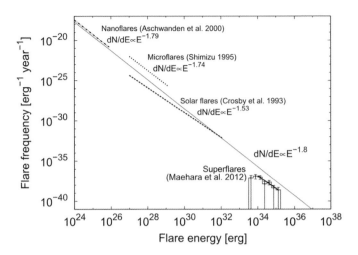

FIGURE 8.2 Comparison between the occurrence frequency of superflares on Sun-like stars and solar flares[+]

always point away from the Sun. The nucleus of a comet is a dusty snowball from the cold depths of the outer Solar System, and when it approaches the Sun, the ice sublimates into a dusty gas that is pushed outwards by solar radiation faster than the comet moves about the Sun. We see the tail because the dust scatters sunlight.

In the case of self-luminous objects like stars, much of the gas is hot and stripped of electrons (i.e. it is ionised), in which case free electrons scatter radiation. Gravity pulls the star together, while thermal pressure and radiation push outwards. In the Sun, thermal pressure dominates, but in the most massive stars, which are also the most luminous ($\sim 10^5 \, L_{\text{sun}}$), radiation pressure is dominant. An upper limit to the luminosity occurs when outward radiation pressure on electrons overwhelms the inward pull of gravity. This was realised in the 1930s by the Cambridge astronomer Arthur Eddington, after whom the limit is now named. Luminous objects persisting for a long time, many years, say, have to radiate below the Eddington limit. Any object seen above the limit should be blowing itself apart and cannot last for long.

A supernova is an example of a super-Eddington event. The core of the Sun is so hot and dense that hydrogen slowly fuses into helium, a process that releases a lot of energy. It will run out of hydrogen in about 5 billion years and the core will collapse, becoming hot enough to fuse helium into carbon and oxygen. Meanwhile, the outer parts of the Sun will swell to engulf the Earth and much of it will blow off into space, leaving the carbon–oxygen core to collapse into an object about the size of the Earth (but with a mass about 60 per cent that of the present Sun). Such an object is known as a white dwarf, which will be the endpoint for the Sun. White dwarfs are made of a very compressed form of matter, known as degenerate matter, which has densities exceeding a million times that of water. Relativity imposes an upper mass limit to white dwarfs of about 1.4 times the mass of the Sun, for a carbon–oxygen composition. This mass limit is named after Subrahmanyan Chandrasekhar, who discovered it while on a ship from India to England, en route to Eddington in Cambridge.

Many stars, however, have a companion star, forming a binary system with the two stars orbiting each other. If they start out close enough, when the outer envelope of the more massive and faster-evolving star swells, then much of that envelope can be captured by the less massive star. Later, the evolution of the second star can transfer matter back onto the first star, which is now a white dwarf. The net result is that the white dwarf can accrete so much mass that it reaches the Chandrasekhar mass. At that point, the object collapses inward, causing fusion of the carbon and oxygen, leading to a massive thermonuclear explosion releasing about a solar mass of newly synthesised iron and radioactive nickel. Over the next few months, the nickel decays to cobalt and then iron, releasing energy as it does so and powering a bright supernova Ia outburst, which can rival a small galaxy for brightness ($\sim 10^9 \ L_{sun}$) (Figure 8.3). About half the iron we see around us and have in our blood has been synthesised in this manner, well before the Solar System formed.

When a star more than eight times the mass of the Sun ends its life, it has an iron core that cannot release more energy through fusion. The core collapses under gravity to densities far beyond those of white dwarfs. The collapse is halted only when the atomic nuclei touch and rebound outwards. The enormous amount of gravitational energy released by the infalling matter heats the bouncing core to such high temperatures that copious numbers of neutrinos are formed. These ghostly particles barely interact with normal matter, but at the extreme

FIGURE 8.3 Type Ia Supernova 1994D (*lower left*) in the outskirts of galaxy NGC 4526[5]

densities in the core their interaction provides an outward pressure that helps to eject the outer envelope of the star into space at tens of thousands of kilometres per second (km/s). This leads to a supernova known as a Type II or core collapse outburst, which is highly luminous but usually less bright than a Type Ia. Type II supernovae are the source of many of the lighter elements such as oxygen, silicon, and sulphur.

There has not been a supernova in our galaxy bright enough to be seen by the naked eye since the time of Johannes Kepler in 1604. There was a relatively nearby one at a distance of 3,000 light years in 1006 that rivalled the full moon in apparent brightness. There has been only one naked eye supernova since Galileo Galilei first used a telescope for astronomical observation in 1609, and that is SN 1987A. This occurred in February 1987 in a neighbouring dwarf galaxy known as the Large Magellanic Cloud, familiar to Southern Sky watchers. Fortunately for me, I was able to see it with binoculars while at the Anglo-Australian Telescope late in 1987, by which time SN 1987A had faded below naked eye visibility.

The collapsed core of most Type II supernovae leaves behind a neutron star or a black hole, depending on whether the original star was less or more than about 25 solar masses, respectively. Neutron stars result from the extremely high densities reached in the collapse when protons and electrons combine to form neutrons. A neutron star is essentially a giant atomic nucleus held together by self-gravity, rather than the strong nuclear force that binds the component particles (protons and neutrons) together in the atomic nuclei around and within us. It is a thousand times smaller than a white dwarf, so it has a density a billion times higher and a radius of about 10 km.

Although neutrons have no electric charge, about 5 per cent of a neutron star consists of protons and electrons, which are, of course, charged. They enable a neutron star to carry a strong magnetic field. Angular momentum conserved in the collapse of the stellar core means that a newly formed neutron star spins very rapidly, at hundreds of times per second. Michael Faraday taught us that moving magnetic fields generate electric fields, which led to the dynamo and the generation of most of our electricity in power stations. A rapidly spinning magnetised neutron star generates such strong electric fields that it rips electrons out of the surface and accelerates them to emit electromagnetic radiation over a wide bandwidth from radio to gamma rays.

Such objects are known as pulsars and were discovered in Cambridge 50 years ago (1967) by Jocelyn Bell and Anthony Hewish. A famous pulsar discovered the following year lies in the expanding supernova remnant known as the Crab Nebula (Figure 8.4) and is seen to pulse 30 times a second. The magnetic field at the surface of that neutron star is estimated to be over a trillion times stronger than the Earth's magnetic field. The energy source for a radio pulsar is the rotation of the neutron

FIGURE 8.4 The expanding debris of SN 1054 in the constellation of Taurus[6]

star. This is confirmed by repeated observations that show the spin rate slowly dropping with time.

Pulsars can be powerful sources of radiation but not more so than massive stars. Some neutron stars have magnetic fields that are a thousand or more times stronger than a typical pulsar like that in the Crab. These are known as magnetars and sometimes lead to very powerful flashes of radiation. The powerful magnetic field provides extreme stress to the solid outer crust of the neutron star. Occasionally, the crust fractures and the magnetic field is rearranged, creating strongly changing fields and intense radiation seen in the gamma-ray band (Figure 8.5). The most extreme flash of radiation, of any wavelength, seen so far was a gamma-ray flash observed by several orbiting satellites in late December 2004. The luminosity of the flash was equivalent to $10^{14} L_{sun}$ or about 1,000 times the radiated power of all the stars in our galaxy – for just 1/8 s. In the 5 minutes after the flash that the object was detected for, pulsations with a period of 7.5 s were detected (Figure 8.6). The gamma-ray flash was intense enough to cause additional ionisation of the Earth's outer atmosphere, which affects the transmission of

FIGURE 8.5 Schematic illustration of the surface layers of a magnetar cracking as the ultrastrong magnetic fields rearrange themselves[7]

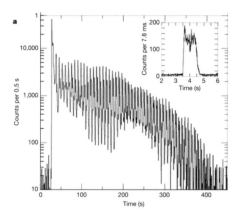

FIGURE 8.6 Hard X-ray lightcurve of the giant flare from the magnetar SGR 1806–20 in the constellation of Sagittarius[8]

some radio waves. The 7.5 s pulsation was picked up from that, too. An explosion happening 50,000 light years away on the other side of our galaxy had a slight, but measurable, transient influence on our atmosphere!

Magnetars in distant galaxies are probably responsible for flashes seen in the radio band and observed as enigmatic fast radio bursts. Initially, only one burst had been seen, and astronomers were beginning to wonder whether it really was cosmic in origin. This suspicion grew when some further radio bursts were found to coincide with the opening of the door of a nearby microwave oven. However, more bursts were later found, including one that repeated in the same position in the sky, meaning it was not due to a catastrophic event like a supernova. The source was then tracked down to a tiny galaxy about 3 billion light years away. It still repeats at times and is likely to be a very young magnetar, less than a few centuries old.

We now turn to black holes, which are responsible for the most luminous persistent objects in the Universe, the quasars. Black holes form in the collapse of massive stars, as mentioned above. Unlike white dwarfs and neutron stars that have upper mass limits of 1.4 and about three solar masses, black holes can have any mass. We suspect that astrophysical black holes have masses that range from about 3 M_{sun} to 10 billion M_{sun} or more. Those with masses above about 100 M_{sun} probably started out with lower masses and grew to their present mass by accretion of gas from the surrounding galaxy.

Massive black holes can also swallow stars whole, if they stray close enough, and if another black hole comes close the two can merge to make a single larger black hole.

A black hole is characterised by only two parameters, mass and spin. They have no other property, sometimes described by the phrase 'black holes have no hair'. The central mass is shielded from our view by an event horizon. Matter and radiation can fall into a black hole but cannot come out; they really are completely black.[9] The radius of the event horizon of a non-spinning black hole (known as the Schwarzschild radius) is 3 km per solar mass, so it is about 30 km for a 10 M_{sun} black hole and 150 million km (the distance between the Earth and Sun) for a 50 million M_{sun} black hole. The radius is reduced if the black hole is spinning and is half the Schwarzschild radius when the spin is maximal.

Black holes grow by swallowing surrounding gas, which usually has some angular momentum. The gas orbits around the black hole, forming a disc within which matter flows inwards and angular momentum is transferred outwards. Gravitational energy is liberated in this process (this is the origin of energy in a hydroelectric power station). It is the most energy-efficient way to release energy from matter. Conversion of matter to energy can be written as $E = \varepsilon Mc^2$, where ε is the efficiency. For chemical processes, epsilon is less than 10^{-10}; for complete nuclear fusion from hydrogen to iron, it is 0.7 per cent; and for accretion onto a non-spinning black hole, it is about 6 per cent. It can rise to about 30 per cent if the black hole spins rapidly. It is this high efficiency that makes accreting black holes so potent and able to produce the most luminous persistent sources of radiation.

The best observational evidence for a black hole emerges from detailed studies of the centre of our galaxy by two groups, one in Munich led by Reinhardt Genzel using telescopes in Chile, the other in Los Angeles led by Andrea Ghez using telescopes in Hawaii. They have made high spatial resolution infrared observations revealing bright massive stars orbiting about an unusual radio source known as Sgr A* (it is in the constellation Sagittarius). Most of the stars in our galaxy, including the Sun, orbit around the centre at about 230 km/s, but those close to Sgr A* orbit at up to 5,000 km/s. Something massive must lurk in Sgr A*. Measuring the velocities and orbital radii of the stars yields a mass of 4 million M_{sun}. The closest orbits constrain its size to less than 10 astronomical units. The only object known to science that can be so massive and yet so small is a massive black hole.

Over the past 20 years it has been discovered that all massive galaxies, and many lesser ones, have a supermassive black hole in their nucleus. The mass of the black hole is typically about 1/500 of the mass of the central stellar bulge of the host galaxy. The mass of the black hole in the giant elliptical galaxy M87 at the centre of the nearby Virgo cluster of galaxies is about 6 billion M_{sun}, and a few other elliptical galaxies have been found to have black hole masses even two to three times higher than M87. They have grown to such enormous masses through accretion of gas at such high rates that they become the most luminous persistent objects in the Universe. The luminosities of such quasars can exceed that of their host galaxy by a factor of 100 or more. They were common when the Universe was about one-fifth to one-third (Figure 8.7) of its present age, i.e. 9 to 11 billion years ago.

FIGURE 8.7 The highly luminous quasar 3C273 lies at a distance of about 2 billion light years in the constellation of Virgo[10]

We have come to realise that quasars had a profound effect on their host galaxies in the sense that the enormous radiation emitted caused much of the gas in galaxies to be ejected, through the action of radiation pressure and winds, so that new star formation was brought to a halt. Blowing the gas away also switched off the fuel for accretion, so the quasar died. This has led to most giant ellipticals now appearing 'red and dead': red because the stars are old and dead because no more star formation is taking place. Star formation now tends to proceed in lower mass disky or irregular galaxies where such feedback has yet to operate.

As matter is funnelled by gravity into black holes, magnetic fields in the gas are wound up, amplified, and pushed out along the rotation axis, creating powerful jets on either side of the black hole. They squirt matter into space at speeds close to that of light, which means radiation from the jets is beamed along that axis. This relativistic aberration of light is so strong that we only see the jet pointing towards us. If it is pointed directly at us, the jet can appear very bright. The first cosmic jet seen was by Heber Curtis in M87, just over 100 years ago. He was looking at a photographic plate and recorded seeing a 'curious thin ray ... connected with the nucleus by a thin line of matter'. Observations of the jet with the Hubble Space Telescope show knots moving along at six times the speed of light. This superluminal motion is a relativistic illusion caused by the jet and its knots moving towards us at almost the speed of light. Such high velocities mean that the jet is very powerful. In some distant objects they are intrinsically much more powerful and are known as blazars and account for the brightest objects seen at those distances.

Jets also power gamma-ray bursts. Discovered in the late 1960s by satellites carrying gamma-ray detectors launched to monitor a test-ban treaty against nuclear weapons in space, they consist of bright flashes of gamma rays appearing at random positions in the sky, about once a day. The energy flux (in gamma rays) is comparable to what we see in the optical band from Venus, which is the brightest persistent object in the sky apart from the moon. Their origin was a major puzzle until one burst was identified with a distant galaxy in the mid-1990s. We now know that the bursts lasting for tens of seconds are due to a spinning black hole forming at the centre of a massive star. Jets, powered by accretion onto the black hole, burrow their way out of the star, producing the burst as they emerge and shock on surrounding matter. The star, meanwhile, explodes, causing a Type II supernova to appear a few days later. The enormous power and beaming of the jets mean that gamma-ray bursts have enabled the discovery of some of the most distant galaxies known.

Short gamma-ray bursts have a different origin, for they occur at the merger of a pair of orbiting neutron stars. The first such system was discovered in the 1970s by Hulse and Taylor and is known as the Binary Pulsar. A pulsar, acting as a very precise clock, in an 8-hour orbit around

FIGURE 8.8 Artist's illustration of the merger of two black holes and the gravitational waves that ripple outwards as they spiral towards each other[11]

another neutron star was a windfall for the study of strong gravity. One very important effect measured was that the two neutron stars are slowly spiralling together at the rate predicted by Einstein in 1916 from general relativity for the emission of gravitational waves. These are ripples in spacetime created by the acceleration of massive objects (Figure 8.8). The rate speeds up as the binary orbit reduces in size until the neutron stars merge in about 100 million years. The magnetic fields of the neutron stars become entwined, funnelling some of the energy of the explosion out in powerful jets. The intense and extremely hot fireball of the explosion is very rich in neutrons that pile into atomic nuclei synthesising the heaviest elements, which are not able to form in core-collapse supernovae. The existence of gold and platinum on Earth is the result of such highly powerful neutron star–neutron star mergers.

Direct detection of gravitational waves requires the most extreme measurements yet made. The passage of gravitational waves from likely cosmic events, such as a neutron star merger, requires a sensitivity to events hundreds of millions of light years away. At the expected rate of mergers of about one per million years per galaxy, the search has to include millions of galaxies. As a gravitational wave sweeps over the

Earth, everything is rhythmically compressed and expanded by one part in 10^{21} in one direction relative to the perpendicular direction. By looking for such a tiny signal using reflected laser beams along evacuated arms 5 km long, in a configuration known as a Michelson Interferometer, the Laser Interferometer Gravitational-Wave Observatory (LIGO) team made their first detection in September 2015. Two interferometers were used, one in Livingston, Louisiana, and the other in Hanford, Washington State (Figure 8.9). The event seen was not of the merger of neutron stars, as expected, but of the merger of two black holes, each about 30 M_{sun} and more massive than any other stellar mass black holes seen to date. The high mass was a bonus, as the signal was very clear in both detectors. During the final merger event, lasting 0.8 s, about 3 M_{sun} of energy was converted into gravitational waves generating a power of 10^{24} L_{sun}. It was the most powerful event ever detected, by a large factor, and amounted to about 8 per cent of the ultimate power!

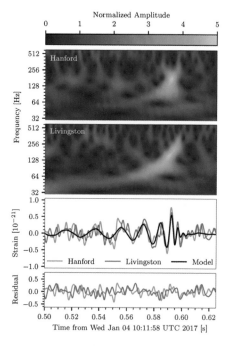

FIGURE 8.9 Third gravitational-wave event detected by LIGO[12]

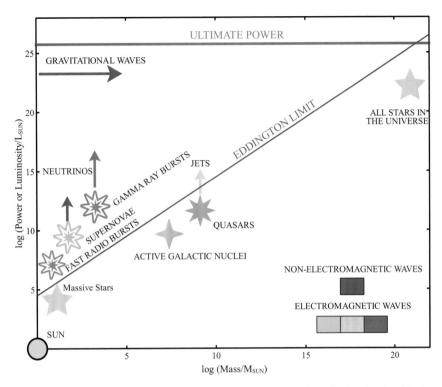

FIGURE 8.10 Logarithm of power in units of the solar luminosity (L_{sun}) plotted against logarithm of mass in units of solar mass (M_{sun})[13]

At the time of writing in late 2017, the detection of gravitational waves has been awarded the 2017 Nobel Prize in Physics (to Rainer Weiss, Barry Barish, and Kip Thorne). Other black hole–black hole mergers have been detected, and in August 2017 the first direct detection of gravitational waves from a neutron star merger was made from a galaxy about 120 million light years away. A simultaneous short gamma-ray burst was detected by two orbiting satellites: the power in electromagnetic radiation was one millionth of that in gravitational waves, yet the information content was much richer. Optical and infrared spectroscopy was possible, and heavy element nucleosynthesis is being confirmed.

We have in this chapter ranged over the extremes of power from solar flares through white dwarfs and neutron stars to supernovae and black hole mergers, covering almost 27 orders of magnitude (see Figure 8.10). The underlying power source is either rotation or gravity, sometimes mediated by magnetic fields. Some extreme events pose a threat to us (solar flares), while many (supernovae and neutron star mergers) create the elements vital for our existence.

Further Reading

K. Blundell, *Black Holes: A Very Short Introduction* (Oxford University Press, 2015).

B. Clegg, *Gravitational Waves: How Einstein's Spacetime Ripples Reveal the Secrets of the Universe* (Icon Books, 2018).

B. Gaensler, *A Guided Tour of the Fastest, Brightest, Hottest, Heaviest, Oldest, and Most Amazing Aspects of Our Universe* (TarcherPerigee, 2012).

Notes and References

1   We briefly derive the simple equation for the maximum or ultimate power that can be released from a mass, M, and radius, R. The largest energy available is the rest–mass energy, $E = Mc^2$. From causality, the shortest timescale on which the energy can be released in an isotropic manner is the light-crossing time (the time light takes to cross the object), which is $t \sim R/c$. The smallest size a mass can have is if it is a spinning black hole, where the size of the event horizon $RBH = GM/c^2$. If we want to see the energy released, it has to be larger than RBH, so $t > GM/c^3$. The net result is that the ultimate power $P = E/t = c^5/G$. (Relativistic effects are important close to black holes and can affect the observed power by a factor of a few.)

2   NASA/Solar Dynamics Observatory, *One Giant Sunspot, 6 Substantial Flares*, www.nasa.gov/content/goddard/one-giant-sunspot-6-substantial-flares/ (accessed 26 July 2018).

3   Low-mass stars do experience flares that outshine them. The nearest star to us, Proxima Centauri, is low mass and in 2016 emitted a flare of energy of $10^{33.5}$ erg, which was so bright that it was almost visible by naked eye on Earth.

4   Data from NASA's Kepler satellite. 1,035 erg superflares are estimated to occur on average about once every 5,000 years on Sun-like stars. Such an event would be hazardous to modern civilisation; K. Shibata, H. Isobe, A. Hillier, et al., 'Can superflares occur on our Sun?', *Publications of the Astronomical Society of Japan*, vol. 65, issue 49 (June, 2013), pp. 1–8. Image reprinted by permission of Oxford University Press. For an essay on this topic from The Darwin Lectures, see J. Wild, 'Foreseeing space weather', in L.W. Sherman and D.A. Feller (eds.), *Foresight* (Cambridge University Press, 2016), pp. 123–36 (the editors).

5 Galaxy NGC 4526 is 50 million light years away, in the constellation of Virgo. Image from the Hubble Space Telescope; NASA/ESA, the Hubble Key Project Team, and the High-Z Supernova Search Team, *Supernova 1994D*, www.spacetelescope.org/images/opo9919i/ (accessed 26 July 2018); for an essay on this topic from The Darwin Lectures, see R. Kirshner, 'Supernovae and stellar catastrophe', in J. Bourriau (ed.), *Understanding Catastrophe* (Cambridge University Press, 1992), p. 5 (the editors).

6 Shown is a composite of X-ray (blue/white), optical (purple), and infrared (pink) images. The pulsar lies at the centre, energising the nebula as it spins down; NASA/CXC/STScI/JPL-Caltech, *Astronomy Picture of the Day: The Crab from Space*, https://apod.nasa.gov/apod/ap180317.html (accessed 26 July 2018).

7 NASA/GSFC Conceptual Image Lab, *Feature: Cosmic Explosion Among the Brightest in Recorded History – Animation 2 – 1st Still*, www.nasa.gov/vision/universe/watchtheskies/swift_nsu_0205.html (accessed 26 July 2018).

8 Seen with NASA's RHESSI satellite on 27 December 2004. In terms of energy flux on Earth, this was the brightest event ever detected from outside the solar system; K. Hurley, S.E. Boggs, D.M. Smith, et al., 'An exceptionally bright flare from SGR 1806–20 and the origins of short-duration γ-ray bursts', *Nature*, vol. 434 (April, 2005), pp. 1098–1103. Image reprinted by permission from Springer Nature.

9 Stephen Hawking showed in the 1970s that black holes do spontaneously emit radiation, now known as Hawking radiation. The luminosity of this process is strongly and inversely dependent on the mass of the black hole. It is completely negligible for the astrophysical black holes described here. For an essay by Hawking from The Darwin Lectures, see S. Hawking, 'The future of the universe', in L. Howe and A. Wain (eds.), *Predicting the Future* (Cambridge University Press, 1993), pp. 8–23 (the editors).

10 The first quasar to be identified was 3C273. A jet of highly energetic particles, ejected from the vicinity of the black hole, is seen as a radial line. The cross shapes centred on the brighter objects are diffraction spikes caused by the support structure of the Hubble Space Telescope; ESA/Hubble & NASA, *Best Image of Bright Quasar 3C273*, www.spacetelescope.org/images/potw1346a/ (accessed 26 July 2018).

11 Such waves are ripples in spacetime; T. Pyle/LIGO-Caltech, *Spiral Dance of Black Holes*, www.ligo.caltech.edu/image/ligo20160615f (accessed 26 July 2018).

12 Note the increase in both amplitude and frequency of the signals up to the merger just before 0.60; MIT/LIGO-Caltech, *Data from the Gravitational Wave signal*, www.ligo.caltech.edu/WA/image/ligo20170601d (accessed 26 July 2018).

13 The Eddington Limit and Ultimate Power are indicated by solid lines. The author would like to thank Peter Kosec for help with this figure.

Index

Index